Circular Economy and Sustainable Development Enterprises

Jiansu Mao · Chunhui Li
Yuansheng Pei · Linyu Xu

Circular Economy and Sustainable Development Enterprises

Springer

Jiansu Mao
School of Environment
Beijing Normal University
Beijing
China

Yuansheng Pei
School of Environment
Beijing Normal University
Beijing
China

Chunhui Li
School of Environment
Beijing Normal University
Beijing
China

Linyu Xu
School of Environment
Beijing Normal University
Beijing
China

The Work was first published in 2016 by China Environment Press with the following title: Circular Economy and Sustainable Development Enterprises.

ISBN 978-981-10-8523-9 ISBN 978-981-10-8524-6 (eBook)
https://doi.org/10.1007/978-981-10-8524-6

Library of Congress Control Number: 2018933514

Translated from the Chinese language edition: *Circular Economy and Sustainable Development Enterprises* by Mao J. S., Xu L. Y., Li C. H., Pei Y. S., © China Environment Press 2016. All Rights Reserved. ISBN: 9787511129345.

Printed on acid-free paper

This Springer imprint is published by Springer Nature
The registered company is Springer Nature Singapore Pte Ltd.
The registered company address is: 152 Beach Road, #21-01/04 Gateway East, Singapore 189721, Singapore

Preface

Circular economy is a certain stage of economy that trying to enclose the material flow in human socio-economic system, and in recent decades, it has become an important measure to cope with resource shortages and environmental deterioration. Under this background, people are making extensive exploration into the theory of circular economy and applying it in an increasingly broader scope. However, the operation of circular economy is intertwined with complex impact from human factors and natural factors, which involves resource science, sociology, economics, environmental sciences, engineering technology, and other disciplines. There are many scientific issues to be discussed, especially the issues like social consumption and economic operation revealed in the implementation of circular economy troubling and restricting the sound development of circular economy. Therefore, it has provided an important opportunity to comb the topics of circular economy in a systematic way and ponder over and explore their resolution.

This teaching material has been compiled based on the course construction of the freshmen seminar of teaching reform project in Beijing Normal University titled "circular economy and sustainable development enterprise" (13-10-10) in response to *Several Suggestions of Ministry of Education on the All-round Improvement of Higher Education Quality* (No. J. G. [2012]4) in a bid to innovate teaching method in education and help talents to fit social needs. Thus the teaching material is written accordingly.

Focusing on the topic of circular economy, the teaching material sets up the basic contents of propelling pattern for sustainable development strategy, basic theory and guarantee system of circular economy, construction and management of sustainable development enterprise from putting forward the problems, theoretical research to practical application. The basic theory of the circular economy features material's flow process and basic law in human socio-economic system with the purpose to provide theoretical direction for the implementation of circular economy. And the guarantee and implementation of circular economy focuses on the explanation of our country's circular economy law to help students to understand the current management of circular economy. Each chapter is guided by scientific research issues and not only pays attention to the knowledge module instruction of related concepts, basic laws, and theories, but also integrates such main links as the basis of topic selection and project establishment, research content and scheme, research achievement expression in the scientific research into group discussion and homework sharing. Our purpose is to cultivate the students' ability of active exploration and making new ground in scientific research, and at the same time, to train them as all-around talents with multidisciplinary knowledge able to explore the development patterns of interdiscipline.

Both the material anthropogenic flow and eco-industry planning are the important contents of circular economy. And related studies had been carefully guided by academician Lu Zhongwu of Northeastern University, academician Thomas E. Graedel of Yale University, and academician Yang Zhifeng of Beijing Normal University, and had been supported by U.S. Henry Luce Foundation, national "973" project, science and technology support and other projects. Here, we would like to express our sincere gratitude! And we also want to express our thanks for the full support of Jiang Pan, Sun Mengying, and other students in the compiling of the teaching materials as well as the support of textbook construction project of Beijing Normal University!

Although the authors of this book wish to demonstrate the multidisciplinary property of circular economy in a systematic way from multiple aspects and make reference to management practice, this book still has lots of mistakes and inadequacies due to the limitation of our ability. We sincerely welcome more readers to give valuable suggestions.

Beijing, China
December 2016

Jiansu Mao
Chunhui Li
Yuansheng Pei
Linyu Xu

Acknowledgements

This work was supported by the National Key Research and Development Program of China ("Ecological security technology of urban agglomeration in the Pearl River Delta-Metabolism simulation & ecological risk forecast and precaution in urban agglomeration," No. 2016YFC0502802) and the Teaching Reformation Program of Beijing Normal University (No. 15-02-08, No. 16-03-08).

The English in this book has been edited by Macmillan Information Consulting Services (Shanghai) Co. Ltd.

Contents

1 Introduction

1.1 Course Background

1.1.1 Why Offer a Research–Discussion Course?

First, offering a research–discussion course is timely. Since the beginning of the twenty-first century, the effect of human activities on the ecological environment has become increasingly obvious, and a significant number of resources and environmental problems demonstrate obvious interaction, periodicity, and complexity. To address this challenge, the current generation should develop solid and rich interdisciplinary and multidisciplinary professional knowledge along with good observation, positive thinking, innovation, and breakthrough skills to support the formation of new ideas, methods, and theories to solve problems. These are the basic scientific research knowledge and skills that researchers should possess.

Second is the need for national talent training. The National Education Reform and Development of Long-term Planning Programs (2010–2020) specifies that reform and innovation is a powerful impetus for education development. The Opinions of Ministry of Education on Improving the Quality of Higher Education in a Comprehensive Way specifies that promoting the overall development of people and adapting to the needs of the community is a measure of the level of talent being trained via basic standard and innovative teaching methods, advocating heuristic, inquiry, discussion and participation in teaching, and promoting interaction between scientific research and teaching, translating scientific research achievements into teaching content in a timely manner. As an important academic training center, Beijing Normal University initiated the freshman seminar project in October 2013 (Teaching through [2013] No. 115). Circular Economy and Sustainable Development Enterprises became one of the first approved freshman seminar projects at Beijing Normal University. This project combines inquiry and research and adopts a thematic research-based teaching model to train students' cognitive and research ability through interactive research and academic discussion with the teachers and scientific research writing on a particular topic. The freshmen

J. Mao et al., *Circular Economy and Sustainable Development Enterprises*,
https://doi.org/10.1007/978-981-10-8524-6_1

seminar is required for all first-year undergraduate students, and the number of elective courses is limited to between 12 and 30. Interdisciplinary topic selection is encouraged. Since 2015, this subject has been included in the general education course at Beijing Normal University.

Third is the need for scientific researchers to have a professional career. Facts show that environmental problems are diverse with different regional distribution, complex formation causes, and a profound evolution process, showing the complexity and dynamics of the interactions between human beings and nature. This occasionally makes it difficult for people with various amounts of knowledge or in existing professions or disciplines to resolve all types of environmental problems between regions. It provides an opportunity for multidisciplinary and cross-disciplinary integration and the promotion of new disciplines and also provides a broad demand for the development of interdisciplinary talents with multidisciplinary backgrounds. Additionally, based on a review of academic papers, dissertations, and research projects, there are many critical problems in existing scientific research, such as a lack of concise content, loose content, or even deviation from the theme. The structures of the papers are confusing, and the chapters, paragraphs, statements, logic, and even grammatical errors are significant. It is hoped that the students' logical thinking and expression ability can be improved through the new seminar and similar mistakes can be effectively avoided in future research work.

Finally, the teaching process is not only the transfer of morality and ethics but also the sharing of human pursuit of a better future. Academician Zhongwu Lu believes that with apologies and qualifications, polite letters and martial arts, with ancient and modern learning, and with combination of Chinese and Western cultures, every student can make his or her contribution to the sustainable development of mankind.

1.1.2 What Is the Significance of Offering This Course?

It is significant to offer this freshmen seminar on Circular Economy and Sustainable Development Enterprise for the following reasons:

(1) Circular Economy and Sustainable Development Enterprise involves various subjects, such as environmental science, resource science, economics, sociology, engineering technology, and management. Put differently, offering this course is conducive to the formation of a multidisciplinary, interdisciplinary system to explore the transfer of multidisciplinary knowledge from the compound talent education model as well as to supplement the lack of existing teaching system discipline.

(2) The discussion approach is utilized for the Circular Economy and Sustainable Development Enterprise seminar. It is of great practical significance to further explore the teaching and research of seminar-type courses, to train the students to actively explore, to explore the quality of scientific research, and to improve independent thinking and innovation. Additionally, this course integrates

theoretical analysis and practical application into teaching. It is to improve students' overall understanding of the circular economy, to compensate for the lack of existing theory and practice, and to improve the ability of students to serve the needs of their motherland.

1.2 General Concept of Teaching and Learning

1.2.1 Why Choose Circular Economy and Sustainable Development Enterprise as a Study Topic?

Circular Economy and Sustainable Development Enterprise is primarily selected as a study topic for two reasons: (1) The strategy for sustainable development is based on worldwide resource shortages and environment degradation, while the circular economy emphasizes material circular flow in human socioeconomic system, so that to conserve natural resources and reduce environmental waste discharge. It has become an important method to promote sustainable development strategy. As the basic unit of economic entities, in the implementation of a circular economy process, enterprises should continue to upgrade and improve their sustainable development, becoming an important part of the development of the circular economy. (2) There have been many unsatisfactory problems in the implementation of the circular economy in recent years. For example, on January 21, 2014, CCTV Topics in Focus reported the adulteration of black water pipes in a factory in Jiangsu as the focus of their adulteration of pipes segment. The factory used recycled waste plastics containing toxic and harmful substances, especially a large number of banned medical waste plastics, to produce water supply pipelines, resulting in adverse incidents of drinking water pollution on human health. In addition, some companies collected vehicles to be scrapped and sold them to rural or remote areas for use, resulting in potential safety problems. Although these actions appear to be the implementation of a circular economy, their results were contrary to the original intention of the recycling economy and must be firmly prohibited by means of management policies and security mechanisms. Thus, there are still many problems to be solved in the implementation of the circular economy, which also provides an important opportunity for the study of related topics.

The circular economy has an artificial circulation flow of material at its core, and artificial circulation is the material circulation process that meets the human need in society's economic system. In the flow process, the material is changed by a series of material form and service functions, as well as the transfer of space and energy that produces the transformation and transfer. The material values change, causing energy flow in the human society economic system. The Circular Economy and Sustainable Development Enterprise seminar involves many fields with interdisciplinary and multidisciplinary features, such as resource science, environmental science, economics, and management. As shown in Fig. 1.1, it can provide a scientific basis for related research in a variety of fields.

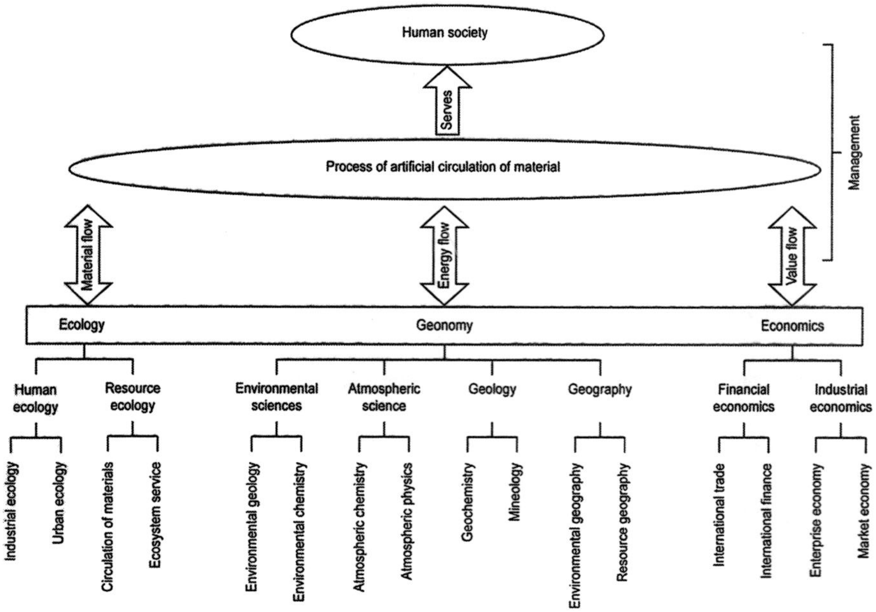

Fig. 1.1 Relationships among subjects involved in the process of material circulation. Reprinted from [1], with permission from Environmental Science & Technology (Chinese)

1.2.2 What Should Be Taught?

Because the Circular Economy and Sustainable Development Enterprise curriculum is based on current, serious environmental problems and sustainable development, resource shortages and deterioration of environmental quality are the major issues that use the strategy of sustainable development as guidance. In many measures, the circular economy and the sustainable development of enterprises are selected as research topics. The relationship between them is shown in Fig. 1.2.

The teaching construction is designed to begin with critical scientific issues, advance to the basic theory of circular economy and security, and finish with the construction and management of enterprise sustainable development in a circular economy. The book is divided into 11 chapters, corresponding to 11 subjects. Each course topic is divided into research topics for students to discuss. The details are presented in Fig. 1.3.

1.2.3 Method to Begin Teaching, Learning and Studying

The entire teaching process includes classroom teaching, topic discussion, and after-class team homework.

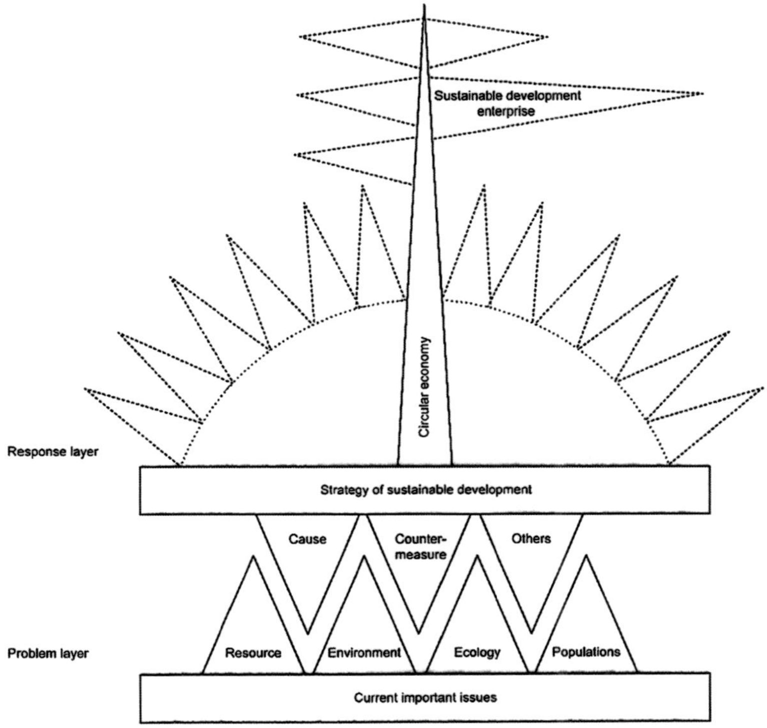

Fig. 1.2 Relationship between Circular Economy and Sustainable Development Enterprise and sustainable development

The classroom teaching addresses two aspects: (1) propose the topic and introduce the content of a chapter and (2) teach by using the knowledge module, which involves the basic knowledge of each chapter as basic concepts, principles, research methods, and laws, demonstrating how to present the essential issues to be addressed, which research methods to use, how to establish research programs, how to perform research, how to display research results, and how to describe the entire process of scientific research. The topic to be discussed by the students should be proposed based on each chapter.

Discussion is the initial step of applying classroom knowledge to scientific research, conducted by the students' scientific team through class discussion. It addresses two portions of Fig. 1.4: topic selection and scientific process analysis. Topic selection focuses on students researching and discussing on the topic. The student team, based on their research interests or expertise, focuses on the appropriate scope of research and selects one specific topic. As Fig. 1.4 shows, in researching the topic of "Is there a shortage of resources?", the students may choose metal resources from land, mineral, forest, and other resources as well as the region China and determine their topic to be Analysis of China's Metal Resources Surplus

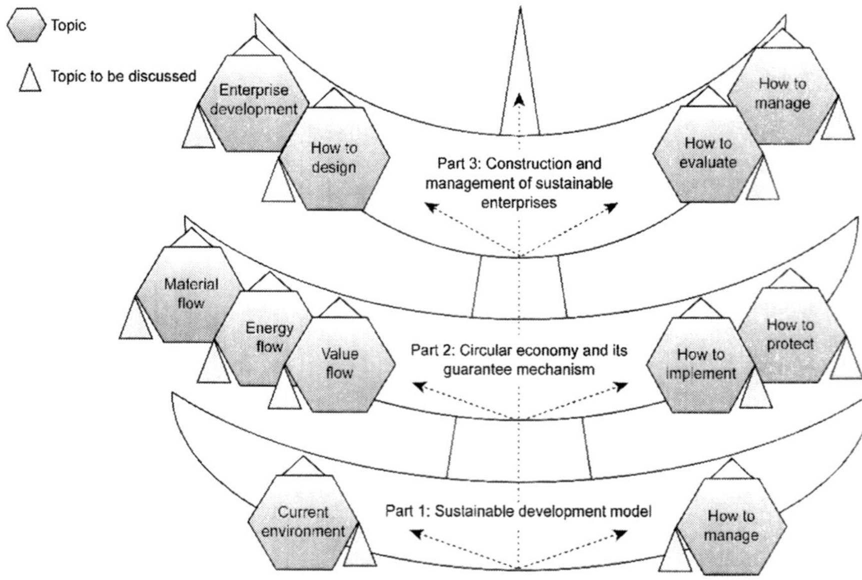

Fig. 1.3 Course topic and content design for Circular Economy and Sustainable Development Enterprise

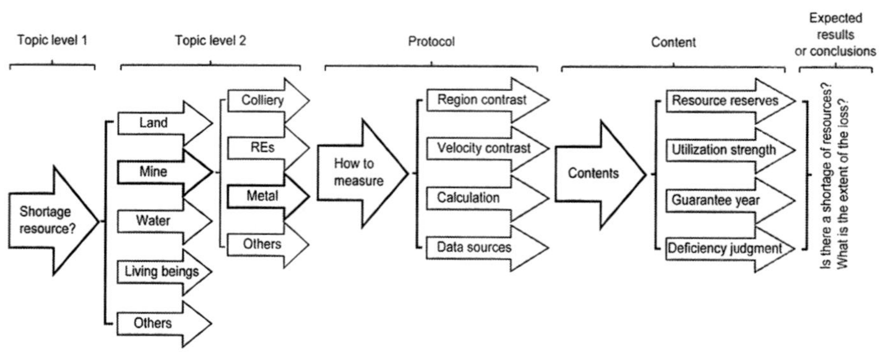

Fig. 1.4 Selection process and analytical framework for the research–discussion process

and Deficiency. During the scientific process analysis of their selected topic, the students discuss the research protocol and method used in the study, the primary research contents, and possible research findings or conclusions.

After-class team homework is not only for students conducting scientific research on selected topics and delivering the scientific research process discussed in class, but it is also the process to test the feasibility of the research protocol and further revise and perfect the scientific work. This after-class team homework is presented as PowerPoint presentations and papers to share with the class.

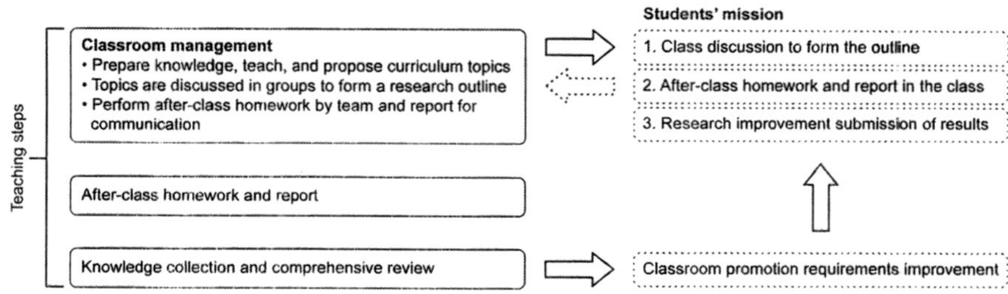

Fig. 1.5 Relationship between teaching and learning in the research–discussion teaching model

The topic discussion and after-class team homework are conducted by a team. Each session, a different team member will organize the class discussion, after-class research, and reporting. Based on this process, each student participates in the group discussion and has the opportunity to organize and report as the representative of the team, comprehensively improving students' thinking, organization, expression, collaboration, and scientific research abilities.

Finally, the teacher and students will all participate in the comment step to correct the after-class team research achievements and revise the issues during the topic discussion. The links between teaching and learning are presented in Fig. 1.5.

1.2.4 What is the Difference from the Traditional Teaching Model?

The discussion approach is utilized for the topic of Circular Economy and Sustainable Development Enterprise, which obviously differs from the compiling and teaching patterns of textbooks in the traditional approach. Each chapter of the textbook corresponds to one of the courses, and each course will focus on a specific subject. This approach requires analysis of the research basis of the subject and the scientific problems that must be solved, identification of the research question of the study, development of a research plan, and the research findings or conclusions. This approach also integrates the basic knowledge displayed in traditional teaching into the entire process of scientific research, from proposing to answering scientific research questions. Additionally, the group discussion and after-school scientific research are performed as scientific research and practice.

Following is the comparison of the teaching processes between the discussion approach and the traditional approach based on an example of global warming.

Under the traditional teaching approach, the process basic concepts-basic theory-basic laws-measures is always used to arrange the teaching content and, in the classroom, the teacher will discuss the greenhouse effect, which gases are greenhouse gases, how the greenhouse gases in the atmosphere cause global warming, the sources of greenhouse gases, how to control greenhouse gas emissions, and so on. Under the discussion approach, the process of observing-question founded-scientific research-problem solved is followed, successively answering the questions concerning how to identify global warming, how global warming is a fact (e.g., the use of glacial history and monitoring records of historic atmospheric temperature), and the effects of global warming on human beings (e.g., sea-level rise and ecosystem deterioration, which must be halted). To determine how to halt global warming, the causes of global warming must be identified; then, related international research and the methods to locate the greenhouse gases that produce greenhouse effects and cause global warming can be introduced. It also can identify the sources of greenhouse gases and how to cope with global warming.

Based on the above comparison, it is easily observed that the traditional approach impacts basic knowledge and the discussion approach is a process of thinking in scientific research. What the students learn from the former is existing research results; however, from the latter, the method to obtain these results is taught to the students, cultivating their scientific research quality.

1.3 Assessment and Evaluation

Course assessment includes regular and final grades, each accounting for 50%. The regular grades focus on the teaching process, evaluating students' scientific research performance in the course of successive issues, including attendance, classroom discussion, and the group report. The teacher proposes the curriculum research quality requirements and standards for evaluation based on progress, and a common score is given by teaching assistants and the students' team. The final grades focus on teaching effectiveness, evaluating the level of scientific research that students can achieve after attending the course and consists of the final group report, graded by the teacher and the team together, and the individual paper, evaluated by the teacher.

Course Resources

Teaching Materials

1. Mao, Jiansu, Linyu Xu, Chunhui Li, and Yuansheng Pei. 2016. *Circular Economy and Sustainable Development Enterprise*. Beijing: China Environmental Science Press.

Recommended and Referenced Literatures

1. Daly, H.E., and J. Farley. 2004. *Ecological Economics: Principles and Applications*. Washington, DC: Island Press.
2. Graedel, T., and J. Howard-Grenville. 2005. *Greening the Industrial Facility: Perspectives, Approaches, and Tools*. New York: Springer.
3. Graedel, T.E., and B.R. Allenby. 2003. *Industrial Ecology*, 2nd ed. New Jersey: Prentice Hall.
4. Ayres, R.U. 1998. *Turning Point and End to the Growth Paradigm*, 2nd ed. New York: St. Martin's Press.

5. Yi, Qian. 2007. *Cleaner Production and Circular Economy: Conception, Methodology and Cases*. Beijing: Tsinghua University Press.
6. Zhongwu, Lu. 2007. *Crossing "Environmental Mountain"-Study on Industrial Ecology [M]*. Beijing: Science Press.
7. Jiansu, Mao, Liang Jing, and Xu Linyu. 2013. The flows of mineral resources to provide human service: concepts, connotation and contents. *Environmental Science & Technology*. 36 (12): 195–201.

Recommended Publications and Websites

1. Ecological Economics http://www.journals.elsevier.com/ecological-economics.
2. Journal of Industrial Ecology http://www.is4ie.org/jie.
3. Resource Conservation & Recycling http://www.journals.elsevier.com/resources-conservation-and-recycling.
4. Journal of Cleaner Production http://www.journals.elsevier.com/journal-of-cleaner-production.
5. Login to the teaching website of "Enterprises of Recycling Economy and Sustainable Development Type". The specific method is as follows: Information Portal of Beijing Normal University\Resource Center\bb Platform\My Course\Enterprises of Recycling Economy and Sustainable Development Type. Students who choose such a course shall use their own student ID and ID card number to log into view. Via query, the course PPT, teaching programs, group assignments and reports, recommended references, course information, interactive pictures and other related contents can be obtained. In addition, the course was published by the International Society for Industrial Ecology (ISIE) in 2014 titled as "New Course on Circular Economy at BNU" as a new lesson in this field. See column "IE News from Around the world" for details, which notes that the course is an international frontier course. URL: http://us2.campaign-archive2.com/?u=d1faa4d588ba0d7d26f41135d&id=d607bea92b&e=fb3943b87.

2 Environmental Issues

2.1 Presentation of Core Issues

The global resource shortages and ecological environment crisis have been increasing since the second half of the twentieth century. In response to this challenge, the strategy of sustainable development was proposed in *Our Common Future* by the World Commission on Environment and Development in 1987. During this period, the circular economy was emerging to reduce the consumption of natural resources and the discharge of environmental pollution based on the circulation of resources in the human socio-economic system. Have the original resource and environmental issues been resolved over the past 20+ years? To what degree were the problems solved? These problems not only affect the implementation but also the next step for further promotion of sustainable development strategy and this course concerning circular economy and sustainable development enterprises. Therefore, the current status of resources and the environment, which is the topic of this chapter, must first be understood.

The course participants are undergraduates from all majors who have seen many environmental problems; however, there is a lack of scientificity and different degrees of understanding of the topic. Therefore, to scientifically answer the questions in this chapter, it will successively present what the environment and environmental problems are, how the emergence of environmental problems is identified and how they are formed via classroom discussion, exercise, and group assignment for good understanding and consolidation.

J. Mao et al., *Circular Economy and Sustainable Development Enterprises*,
https://doi.org/10.1007/978-981-10-8524-6_2

2.2 What Are the Environment and Environmental Problems?

2.2.1 What Is the Environment?

We may have had the experience of breeding. Using fish farming as an example, the fish tank, water, aquatic plants, food, and so on, which can be called the environment, are necessary for the growth and survival of the fish. Similarly for growing flowers, the flowerpot, soil, as well as the added water and nutrients, which can be called the environment of the flowers, are the necessary conditions for the growth and survival of flowers. It can be seen that the "environment" is relative to a central object (also known as the "body").

The environment is an external, human-centered world in the Circular Economy and Sustainable Development Enterprises course, which is a complex of material conditions for human survival and development. The "Environmental Protection Law of the People's Republic of China" stipulates that environment is a combination of various natural and artificially transformed natural factors that affect human survival and development, including atmosphere, water, ocean, land, minerals, forests, grasslands, wildlife, natural and cultural relics, natural reserves, scenic areas, cities and country. Obviously, all the contents included in the concept of "environment" are necessary conditions of human normal life, production, and survival.

2.2.2 What Are the Environmental Problems?

Because the environment primarily serves human beings, environmental problems occur when environmental factors that are not conducive to human survival appear. For a more scientific definition, environmental problems refer to the phenomena that are not conducive to human survival and social development in the global or regional environment caused by natural forces or human activities. Although environmental changes caused by natural forces, such as volcanic eruptions, earthquakes, floods, droughts, and landslides are usually beyond human intervention, recent studies show that with the increasing intensity of human activities and accumulation over time, human activity has become an important factor in some natural disasters (i.e., drought in some areas implies improper land use and other factors), which will also be included in this course. In summary, the environmental problems discussed in this course refer to all types of problems caused by human activities that are not conducive to human survival and development.

In general, based on different objects, there are three basic types of environmental problems, including resource shortage or depletion, environmental pollution, and ecological destruction. Resource shortage or depletion is observed when the natural resource supply has difficulty meeting the needs of human production and life, which affects development of the national economy. It primarily includes the shortage of water, land resources, energy, and mineral resources. For example, traditional business failures or transformation due to the shortage of mineral resources in some areas; timing of the power supply and limited power due to the lack of energy in some areas; and adoption of the South-to-North Water Diversion in North China to satisfy water demand due to the water shortage. All these problems are the result of unreasonable exploitation and utilization of natural resources. Environmental pollution is a phenomenon that leads to changes in the structure and function of the entire environmental system due to diffusion, migration, and transformation of harmful substances into the environment by human activities, which are not conducive to the survival and development of human and other biological beings. Environmental pollution can be further divided into the pollution of water, air, soil, and noise. For example, the haze we are exposed to contains excessive toxic and harmful substances due to changes in atmospheric composition, which seriously affects human health and life quality. Ecological destruction refers to the environmental problems caused by human activities on natural ecosystems, resulting in the change of ecosystem structure, and the decline of production capacity. It includes biodiversity reduction, forest and pasture vegetation destruction, soil erosion and other issues. For example, Baiyang Lake Wetland, the Kidney of North China, is shrinking and in danger of disappearing.

In addition, different types of environmental problems are interrelated and interact. For example, the water shortage is an environmental problem that not only indicates resource shortage but also ecological destruction. The destruction of the ecological environment will reduce the ability of nature to absorb and digest pollutants, which leads to further aggravated environmental pollution.

Although different, these environmental problems all affect the survival and development of human beings from different perspectives, which requires scientific measures.

2.3 Methods to Identify the Emergence of Environmental Problems

Different environment types provide different service functions and support for human beings. Therefore, it is difficult to judge whether they affect the normal existence and development of human beings. We identify environmental problems, using resource shortage and environmental pollution as examples.

2.3.1 Example of Identifying a Resource Shortage

Resource shortage identification usually refers to a phenomenon of the hard resource supply that supports human life and production as the environmental problem of resource shortage or depletion. Therefore, one of the effective methods to identify a resource shortage is to compare the relationship between resource supply and human demand. It is considered a resource shortage when the resource supply is lower than human demand, which will affect the normal existence and life of human beings. It can also be the comparison of the relationship between resource stock and human consumption, similar to the relationship between a fund source or bank deposits and living expenses. The amount of resource may be abundant even if there is a limited amount, it is not used. Therefore, a resource shortage is relative to human demand and consumption.

Generally, resources can be divided into non-renewable and renewable resources.

The replenishing of non-renewable resources is very slow, such as the formation of mineral resources, which requires tens of thousands or even millions of years; however, they are consumed very rapidly. If half of the mineral reserves have been consumed in the past two to three hundred years, the mineralization rate is almost zero by contrast. It can be considered that the total amount of resources does not change with mineralization and a certain amount is consumed by human beings yearly, which leads to an annual decrease in the mineral resources. In mining management, the ratio of the stockpile and annual exploitation amount of a certain mineral resource is usually defined as the guaranteed period. The longer the guaranteed period is, the more abundant the resource; however, the shorter the guaranteed period is, the less abundant the resource, meaning that alternative resources must be located as soon as possible; otherwise, human production and life demands will be seriously affected.

For renewable resources, such as grassland and woodland, human use of the resources should not exceed their replenishment rate. For example, if grazing grassland region can support 100 sheep, but it is actually populated by 150 sheep, the grassland will eventually be destroyed because the sheep, with their demand for food, may eat the shoots that have not yet grown or even roots. This can lead to overgrazing and, combined with passive grassland reduction, can result in food shortages.

2.3.2 Example of Identifying Environmental Pollution

Environmental pollution generally refers to a high content of pollutants that will affect the normal function of the environment. For example, arable land is primarily used for crops to meet human diet needs. However, in recent years, the heavy metal content in the soil has increased in some areas. We know that heavy metals are harmful to the human body; for example, lead can damage the nervous system and cadmium can harm the skeletal system. Therefore, allowable concentration standards for heavy metals in different environmental categories that do not affect human health are established by countries and are simple to apply. They can be identified as environmental problems when the actual pollutant concentration in the environment is greater than the standard defined by the state.

The services provided to human beings are different for various types of environments and, therefore, so are the required quality standards. Related standards have been established based on different environments, and the varieties, testing methods, and concentration limits for environmental pollutants have been defined by the International Standardization Organization and countries to guarantee environment quality and meet the requirements of human survival. Examples of standards include the *Ambient Air Quality Standards (GB 3095–2012), Surface Water Environment Quality Standard (GB 3838–2002), Soil Environment Quality Standard (GB 15618–1995)* and *Surface Water and Wastewater Monitoring Technical Specifications (HJ/T 91–2002).*

There is another method to assess the degree of environmental pollution called the rate contrast. It is the contrast of the relative rate difference between anthropogenic emission and natural circulation. When the anthropogenic emission rate is lower than natural circulation, it means that the emissions from human activities will be transferred into the environment with natural circulation, which will not affect the normal function of the natural system. However, when the emission rate of human activities is higher than natural circulation, the excess pollutants will accumulate, leading to a rise of content in the environment. This will affect quality of service subject if the concentration is allowed to increase beyond the allowed range.

Other methods have been established by the state for easy application, such as emission permits in some regions. It is considered a pollution problem when the actual emissions exceed the specified quantities.

2.4 How Environmental Problems Are Formed

Because environmental problems are caused by human activities, the formation of environmental problems must be related to the relationship between human beings and the environment. Accordingly, what is the relationship between human beings and the environment? The following sections provide a qualitative analysis.

2.4.1 The Relationship Between Human Beings and the Environment

From the qualitative point of view, human beings are the product of the continuous evolution of the surface ecosystem, interact with their living environment and restrict each other with a special relationship that is both a contradictory and unified existence.

Substance and energy must be obtained from the natural environment for human life and production. It forms the flow of human society and the economic multiplexed system due to subjective needs and purposeful activities, which disturb the natural flow of substances and energy and inevitably contradict the formation and transformation of material and energy in the natural environment, restricted by the ecological environment. This is the contradictory existence.

This requires human beings to fully understand the natural environment and arrange production and living in accordance with the natural law of evolution; otherwise, they will be punished by the environment.

Uniform refers to the most active ecological components of natural ecological systems. The natural ecological system carries human beings, with humans as the most active ecological group. Human survival and development depend on the ecological environment. Human activities are the interaction between human beings and the environment; the process of material and energy exchange. Mankind is not only the product of the long-term evolution of the natural environment, but is also the creator of the environment.

2.4.2 Environmental Load and Carrying Capacity

Everyone has the experience of lifting, and whether it is successful depends on the differences between the human's strength and the weight of object. As the carrier of human beings, the natural environment system also has this problem. In environmental science, the scale of the population and economy that environmental resources can accommodate at a specific level is called the environmental carrying capacity, under the premise of a relatively stable environmental system maintained over a certain period. It is the support to the development of human society. The disturbance to the environment system produced by human beings under certain conditions of population, economic scale, and technology is called an environmental load, which is the pressure of human activities on the natural environment. When the pressure exerted by human beings on the environment is less than the environment capacity, the environment will remain stable; when the environmental load is equal to the bearing capacity, the environment system is in critical state; and when the environmental load is greater than the bearing capacity, the environmental system will likely deteriorate, or even suffer destruction. In Fig. 2.1, P represents the human pressure on the environment and C represents the environmental carrying capacity.

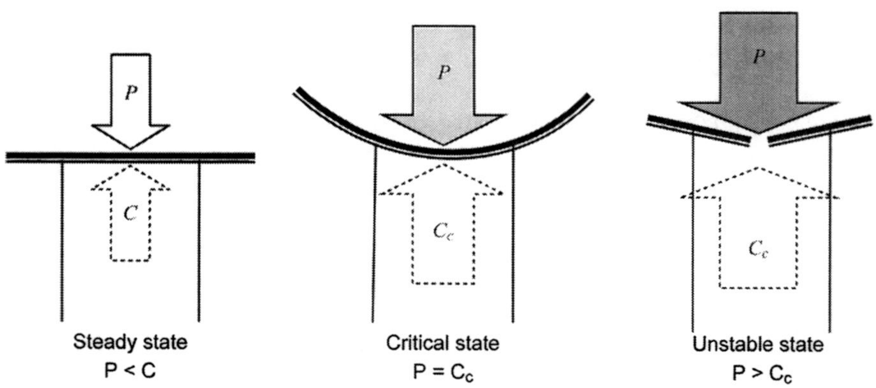

Fig. 2.1 Comparison of environmental pressure and carrying capacity

It is important to ensure the stability of the natural environment and function by controlling the human environmental load to not exceed the carrying capacity. The majority of existing environmental problems are a result of the conflict between the level of human economic activities and the environmental carrying capacity; therefore, the environmental carrying capacity can be used to measure the coordination degree between human societies' economy and the environment. Because the environmental carrying capacity of different regions can be quite different during different periods, human beings can improve the regional environmental carrying capacity by changing the mode of economic growth, improving the technology level, and increasing environmental governance.

2.4.3 The Essence of Environmental Problems

Everything possible is undertaken to obtain resources from nature to create living conditions for human survival and development. When human behavior does not exceed the allowed threshold, the ecosystem balance can be restored after mediating the interference through the self-regulation mechanism of nature, which fosters harmony between man and nature. For example, human beings primarily existed on farming in ancient times and could live in harmony with nature because the transformation of nature was only to grow crops or domesticate animals, which has little effect on nature. However, after the industrial revolution, with the great improvement of production capacity, the transformation and intervention on nature by human beings are increasingly strengthened, even beyond the bearable limits of natural systems, and a series of environmental problems have emerged because the balance can no longer be restored through the self-adjustment mechanism. Overall, environmental problems are caused by human beings from their own needs, such as careless and excessive access to limited resources and wanton emission of metabolites into the environment regardless of the consequences. This "grab" and "emission" strongly interfere with the natural circulation and energy flow of the surface system, breaking the original ecological balance, without fully understanding, and having concern for the interdependence between human beings and the environment. Simultaneously, with the development of human beings, the population scale and demand are increasing, which leads to an increasing trend of claiming and releasing emissions to nature. The contradiction between supply and demand is becoming increasingly obvious with the increase of the human environmental load and the decline of the environmental carrying capacity, which is the root cause of environmental problems.

Currently, the problem confronting human beings is that the environment is increasingly deteriorating and the speed of resource exploitation is faster than its regeneration or generation of a substitute. The amount of waste discharged into the environment exceeds the environmental capacity and self-purification ability. Therefore, the essence of the environmental problem is the maladjustment of the relationship between human beings and the environment, which leads to the destruction of the ecological system balance. It is caused by the incongruous development of the population, economy, society, and environment and by the ignorance about and disrespect of the laws of nature. Thus, it is possible to achieve a real sense of harmony between human beings and the environment only by fully understanding the dependency, fundamentally changing the demand mode, abandoning the egoism of blind nature demand to meet our personal and short-term interests, and changing the manner of production and life.

Classroom Discussions and Assignments

Develop a study of current environmental problems at home and abroad through classroom discussion. Consider their extent. This will clarify the current resource situation and environmental problems, a prerequisite for beginning this course, and also provide an example of a potential class discussion.

Discussion Topics

To understand the current situation of resources and the environment, please choose one of the following topics. Select a typical environmental problem as the discussion topic for the group and explain the reason for the choice. Environmental problems have been divided into resource shortage, environmental pollution, and ecological destruction. A discussion will be performed within the group, including how to study and the content.

1. What are the current levels of resources?
2. What is the quality of the ecological environment?
3. What is the status of environmental pollution?

Example of Topic Selection

The topic cannot address everything in the practice of scientific research due to a wide range of resources, a diversified ecosystem, and environmental factors. Otherwise, the research will be general, it will be difficult to be in-depth and it cannot produce valuable results. Therefore, screening issues are the first step. The selection should be implemented from beginning to end, focusing on a specific problem for in-depth research of the selected topic.

Topics can be screened by employing the following method: ① classify the topic based on extension of the keywords in the topic and select one for research, i.e., "resources" can be divided into non-renewable resources, renewable resources, and constant resources based on their renewability. Select "non-renewable resources" for research; ② select a topic based on the location and time period for the study, such as global or China, or a province for regional study for a specific year, age, or century; ③ study the contribution of greenhouse gas emissions to global warming based on the causes and mechanisms of the problem, such as the Intergovernmental Panel on Climate Change (IPCC); and ④ select typical and representative problems based on their degree, such as mineral resources. In addition, the focal point of the project, required completion time, financial support, scientific research strength of the possible team, and other factors may also impact the choice of topic.

The result of the topic selection will be a study of a typical problem within a specific location and time, based on the above influencing factors.

A discussion topic for the Circular Economy and Sustainable Development Enterprise seminar should be as limited as possible and focused on a specific issue to allow students to learn and apply research concepts. This also considers that the students are new undergraduates, this is their initial contact with scientific research work, they have extremely limited time and they do not have the required professional knowledge. For the three core topics in this chapter, the recommendations for classroom discussion are as follows:

1. Study the current situation of the coal mine resources in Shanxi;
2. Study the ecological environment in the Baiyangdian Lake Basin; and
3. Study heavy metal soil pollution in the Jinzhou farming area.

Using the third topic as an example, the selection process is demonstrated in Fig. 2.2. When considering that Jinzhou is one of the rice producing areas in China, the heavy metal content in the soil will further affect rice quality and human health; therefore, the status of heavy metal soil pollution in the farming area is chosen as the research topic.

Example of Research Protocol Selection

Different topics have different research protocols, primarily depending on the inherent rule of the problem to be studied, and the scientific method employed to reveal the rule. It usually includes the selection of the research object, the schedule and location, the methodology, the technical route, the data acquisition and calculation, and so on.

(I) Method of study: material cycle rate comparison

In the topics of this lecture, it is emphasized in sustainable development that for renewable resources, resource consumption should not exceed the regeneration rate, and for non-renewable resources, the consumption rate should not exceed the development speed of alternative resources. In terms of waste emissions, the pollutant emission rate should not exceed the biogeochemical cycle rate. Therefore, the

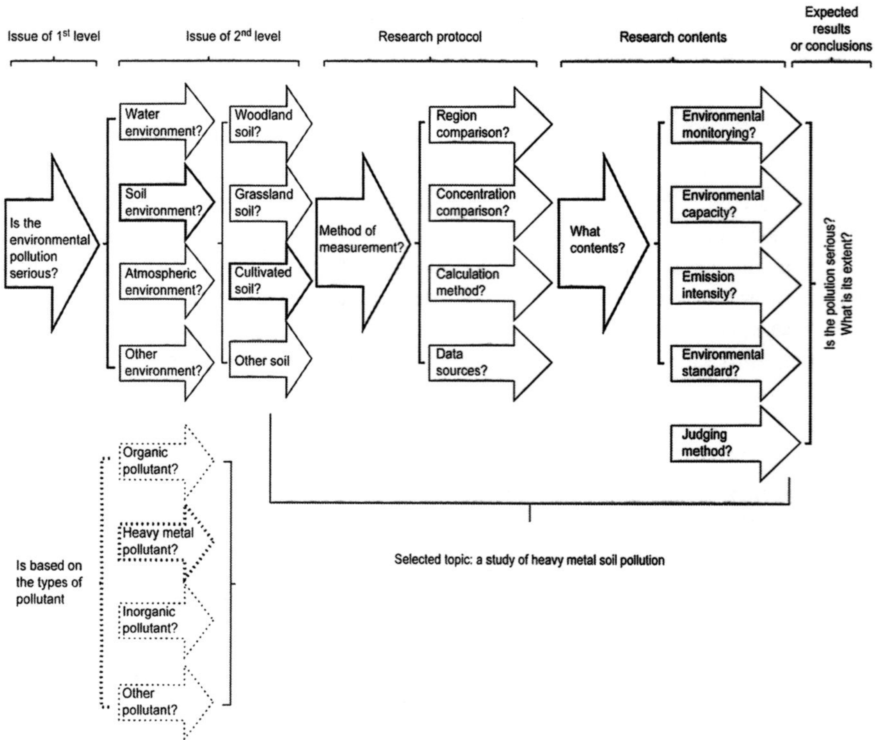

Fig. 2.2 Selection of class topics and the analytical framework for discussion

rate comparison method can be used to understand the resource and environment status.

Karlsson [5] provides a useful results from a comparison between the artificial and natural circulation rates for several important metals, such as Al, Mn, Mo, Pb, and so on. It is observed from his research that, the artificial circulation rate.

In addition, it is common to understand the soil pollution degree by monitoring the actual concentration of pollutants in the soil, compared with the national standards of the pollutants' limits.

(II) Examples of data sources and data processing methods

Effective and sufficient data, which are the prerequisite and basis for effective research results, have the same importance as research methods in a scientific study. For different research topics, the questions to be answered are different, requiring different data; therefore, the data sources and calculation methods are not the same.

Table 2.1 Several guaranteed period estimates for the world's metals as of 2014

Metal symbol	Mineral production (million tons/a)	Reserves/million tons	Guaranteed period/a
Zn	1010	22,000	22
Cu	1490	47,000	31
Ni	150	6200	40
Pb	328	6700	21
Ag	2.03	27	13

Data from: http://minerals.usgs.gov/minerals/

When studying global or national status, the three data values for reserve quantity, consumption level, and guaranteed period for all types of metal mineral resources must be known. The guaranteed period is usually estimated based on the ratio of reserve quantity and annual mineral output. Therefore, the first two are data that must be obtained from a statistical report, and the third, the guaranteed period, can be estimated based on the first two values. The surplus degree of different metal resources can be determined from the guaranteed period; thus, statistical reports from the United States Geological Survey (USGS), the China Geological Survey (CGS), and other land and resource management departments will become an important source of data. Several important Web sites closely related to the three topics discussed in this chapter are listed below.

China Geological Survey (CGS): http://www.cgs.gov.cn/
The Central Peoples Government: http://www.gov.cn/
The Ministry of Environmental Protection: http://www.zhb.gov.cn/
Ministry of Water Resources: http://www.mwr.gov.cn/
The Ministry of Land and Resources: http://www.mlr.gov.cn/
United States Geological Survey (USGS): http://minerals.usgs.gov/minerals/pubs/commodity
National Environmental Protection Agency: https://www3.epa.gov/
The National Energy Administration: http://www.nea.gov.cn
China Statistical Yearbook: http://www.stats.gov.cn/tjsj/ndsj/

Example of Research Results: Guaranteed Periods of Several Important Metals

As critical raw materials for industrial products, metal resources act as the basis of national economy and play an important role in modern industry. The lifespan of the resources can be estimated based on the understanding of resource reserves in the earth's crust and annual mineral quantity, which reflects the surplus degree of different metal resources, as shown in Table 2.1.

The situation of reserves and guarantees in different countries and regions can be analyzed by using a similar method to understand the regional differences in the distribution of resources.

Examples of Classroom Discussion Results

The discussion results can be recorded in the form of a "discussion outline" and table and shared in class by the group representatives with the assistance of other team members and teachers, allowing for further improvement, as shown in Table 2.2.

Table 2.2 Example of the record of key points of discussion results

No.	Item	Key points	Improvement points	Remark
1	Topic name	Study of soil pollution in Jinzhou	Focus on "farming area" and "heavy metal"	Alternative farming areas in Tianjin
2	Primary selection rationale	Desiring to know the quality of the soil of a classmate's hometown	Jinzhou is an important grain producing area in China and primarily grows rice. The soil quality will affect the food and the accumulation of heavy metals in rice has a risk for human health	
3	Proposed research protocol	Field sampling, testing a variety of pollutants compared with the national standard limits	The study scope is narrowed to "farming area" and pollutant is focused on "heavy metals" with four metals (Cr, Cd, Pb, and Cu) selected	
4	Primary research contents	Quality requirements, pollutant testing, comparative analysis, improvement suggestions for the soils of different functional areas of Jinzhou	Supplement: separate testing of the soil in some farming areas, quality comparison of each farming area	
5	Expected results or conclusions	Identify the soil areas with good quality, serious pollution, and exceeding degree	Supplement: quality grading of soil in different farming areas	
6	Team members and division of labor	Student Li assumes overall responsibility, student Wang is responsible for field sampling and testing, student Zhang is responsible for studying the national standards	Supplement: Student Tian is responsible for data collation and calculation and the summary of team contact information	

Based on the discussion results in this class, the team members independently perform research work on selected topics and organize his or her results into a PowerPoint presentation.

Group Work Tips

Please consider the following suggestions during the initial stage of your scientific research:

1. Avoid broad and extensive themes, big topics, and wide coverage;
2. Reports must be logical and clearly structured, focused, contain the scientific methodology used, and be concise and comprehensive. The use of charts and tables is encouraged to present the results; and
3. Reports should last no longer than approximately 10 min.

Possible Conclusions of Group Work

The overall environmental quality situation is very disturbing due to the shortage of various important natural resources. Concurrently, environmental pollution in many areas is still very serious and has even intensified, which threatens human health; overall improvement is urgently needed.

Recommended and Referenced Literatures

1. The United Nations World Commission on Environment and Development. 1987. *Our common future*. Oxford: Oxford University Press.
2. Yang Zhifeng, Liu Jingling. (2010). *Introduction of environmental science*. 2nd Version. Beijing: Higher Education Press.
3. Lin, A., Hu Jiangjun, and Zhang Ling. 2005. *Resource environment and sustainable development*. Wuhan: Wuhan University Press.
4. Qian Yi, Tang Xiaoyan. (2010). *Environmental protection and sustainable development*. 2nd Version, Beijing: Higher Education Press.
5. Karlsson S. (1999). Closing the technospheric flows of toxic metals-modeling lead losses from a LAB system for Sweden. *Journal of Industrial Ecology* 3(1):23–40.

3 Sustainable Development Management

3.1 Presentation of Core Issues

Because many natural resources are limited and the environmental quality of many areas is poor, environmental management is still very necessary. However, how should environmental management be performed? Since the middle of the twentieth century, this issue has been deeply discussed. Particularly, the World Commission on Environment and Development proposed the strategy of sustainable development in an article from 1987, *Our Common Future*. This was a significant milestone, which clearly establishes environmental management objectives and many strategic management measures. More than 20 years have passed—what can we learn from past experience? Facing the resource and environmental issues that have not been satisfactorily resolved, the question remains of how we should move to the next step of management. This is the subject to be discussed in this chapter.

To seek the answer to this course topic, first, we should reflect on history, reviewing the strategy of sustainable development and management. Then, based on exploring the quantitative relationship between human beings and the environment, we can discuss how to perform scientific management, the propulsion measures of sustainable development management, and how the circular economy is incorporated. Additionally, supplemented by classroom discussions and group assignments, students' application skills can be enhanced.

3.2 How Sustainable Development Was Historically Managed

3.2.1 What Is Sustainable Development?

In 1987, the Norwegian Prime Minister, Mrs. Brundtland, formally proposed the definition of sustainable development as "development which meets the needs of

J. Mao et al., *Circular Economy and Sustainable Development Enterprises*,
https://doi.org/10.1007/978-981-10-8524-6_3

current generations without compromising the ability of future generations to meet their own needs." This concept emphasizes the following three points. ① From a management perspective, what is "development"? Sustainable development is different from the society growth. Traditionally, the meaning of "growth" reflects economy, welfare, and other aspects of "expansion," but the use of the word "development" emphasizes the improvement of the "quality" of these elements. For example, although we live in a more spacious house, the surrounding environment is messy, and we have to drink poor water and breathe contaminated air, which is the "growth," but cannot be called "development." Natural resources and the environment are the basis for the survival of mankind. Sustainable development requires not only the coordinated development of human and external resources and the environment but also the coordinated development of human society and the economy. ② From the perspective of human society, what "demands" should we maintain? Are demands to meet the basic needs of fundamental necessities or to pursue luxury and extravagant waste? Different "needs" mean different consumption patterns, which also may produce a different environmental load, affecting the relationship between human beings and the environment. For example, in winter, we wear sweaters in indoor places that have a high temperature; however, in summer, when the indoor temperature is low, we also wear sweaters, which not only affect human health but also cause a significant amount of wasted energy. Over time, this is likely to exacerbate the degree of energy shortages, affecting the normal needs of future generations. ③ From a technical point of view, how do human beings meet their needs? Human demand is an objective manifestation of human subjective desire, such as houses to meet the housing demand and clothing for defending against the cold and concealing the body. The realization of the object requires a human throughout the production process, processing, manufacturing, and transforming the natural resources into the products or services that have a specific function to satisfy specific human needs. It can be seen that human satisfaction with their needs is the process of the interaction between human activities and the natural environment, reflecting the relationship between human beings and the environment. In short, we need careful planning and reasonable arrangements to achieve sustainable development. While continually making progress in economy, culture, and technology, we should maintain the desired level of environmental load to achieve coordinated development between mankind and the environment.

3.2.2 How to Establish Overall Management Objectives

(I) An Example of Overall Objectives of Quantitative Management: the Tenfold Growth Target

To perform sustainable development management more scientifically and effectively, a significant amount of research was performed in the 1990s. In 1997, the famous "*Carnoules' Statement*" was published, which explicitly stated that "A tenfold leap in energy and resource efficiency in one generation" which is a noble goal of a tenfold improvement in the efficiency of using resources, energy, and other substances. This quantified objective is based on the IPAT equation proposed by Prof. Ehrlich and Holdren, well-known demographers at Stanford University in 1971. Following is the estimation process based on the IPAT equation:

$$\text{Environment Impact} = \text{Population} \times \text{Affluence} \times \text{Technology} \tag{3.1a}$$

Which can be translated to the following:

$$\text{Environmental Impact} = \text{Population} \times \text{Welfare} \times \text{Technology} \tag{3.1b}$$

Using the first letters of the English words, the above equation can be simplified to the following:

$$I = P \times A \times T \tag{3.1c}$$

In the above equations, "environmental impact" refers to the impact of a research subject system on the external environment. For example, the human planting process will reclaim land and change the regional species structure and geomorphology. Building houses will require the exploitation of natural resources such as minerals, sand, and wood for the production of building materials. In application, the "environmental load" can be used to quantify the quantity of natural resources or the amount of environmental pollutants discharged. As mentioned in Chap. 2, approximately ten million tons of zinc ore are produced annually in the world. According to statistics, China currently consumes approximately four billion tons of standard coal and discharges 70 billion tons of industrial wastewater each year. This reflects the human socio-economic system having a significant impact on the natural environment.

The right side of the equation includes three variables: The first variable is the population, expressed in terms of population and primarily affected by birth, death, and migration. Based on statistics, the population continues to grow. The second variable is social welfare. In the estimation, economic-level indicators are used to indicate per capita GDP, which is primarily influenced by factors such as social, economic, technological development, and development stages. It is predicted that the economic scale continues to grow. The third variable is technology, which reflects the technical means and technical levels adopted by human beings during the process of acquiring social welfare. If the per capita GDP is used to represent the welfare, the "technology" is the environmental impact of the unit GDP, based on the principle of consistency of the physical dimension of the mathematical equation. This indicator is primarily affected by human activities and nature, including industrial structure, technical conditions.

The IPAT equation quantitatively reflects the relationship between social and economic development and the environment in a given region, which is the basic equation to be followed in environmental management. The equation is also called the master equation of the relationship between mankind and the environment.

At the end of the twentieth century, Friedrich Schmidt-Bleek organized an international club with ten members. The organization initially estimated that the impact level of the global environment has exceeded the earth's carrying level by approximately two times, while it predicted the world population will double and per capita GDP will grow three to five times in the next 50 years, meaning the global economy will grow six to ten times. This is the famous Carnoules Declaration. Following this, the organization also published a number of strategic management masterpieces (See http://www.factor10-institute.org/publications.html).

(II) Examples of Qualitative Management Objectives

Due to the geographical distribution of natural resources as well as regional differences in human factors such as society, economy, and technology, the environmental problems faced by the world have their own characteristics, and the relationship between human society and the local environment is different, so management objectives will vary from location to location. Here, the qualitative management objectives of developed and developing countries are reviewed separately.

1. The Environmental Management Goals of Developed Countries

Developed countries have completed the industrialization process. During their industrialization process, they experienced economic growth as well as rapid growth of the environmental load. The relationship between the growth and the environmental load is expressed by Graedel and Allenby in the book of *Industrial ecology* as a reversed U-shape, in which the environmental load increases with the growth first and then decreases when the growth reaches a certain level. The process of development, where the abscissa is the "development situation," has a broader meaning for economic conditions and can be expressed not only by economic growth but also by the human development index (HDI) or other indicators. The ordinate is "resource consumption," emphasizing the source of environmental load. The development process of developed countries can be divided into the following three stages: The industrialization stage in which the environmental load is rising; the vigorously remedial stage in which the environmental load increases at a slower rate, reaching the vertex and gradually declining; and the prospect stage (not yet fully realized). The environmental load continues to decline until it is very low, reaching the desired level.

The focus of environmental management in developed countries is to undertake environmental improvement tasks, with the assistance of technological development and progress, acquiring the important measures and ways that can effectively reduce the overall environmental load.

2. Environmental management objectives of developing countries

The situation in developing countries is significantly different from that of developed countries, which is manifested in economic development, changes in the pace

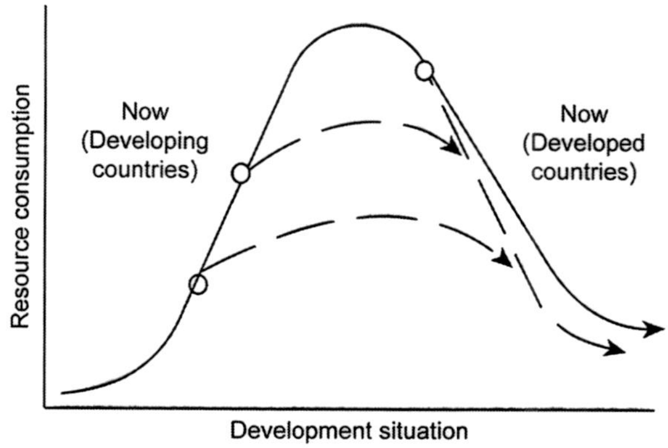

Fig. 3.1 New industrialization road of developing countries: going through the hillside of the "Environmental Mountain." *Source* Lu and Mao [5]

of development and economic structure. If they continue to follow the development path that Western developed countries have experienced, they will inevitably further aggravate the deterioration of global environmental quality. Therefore, we must earnestly learn from developed countries' experiences and not repeat their mistakes.

Lu Zhongwu, an academician of the Chinese Academy of Engineering, vividly described the reversed U-shape curve that describes the change of environmental impact with human development for developed countries as an "environmental mountain" and described the human development process as a mountain of activity. On the basis of the study of the relationship between economic growth and environmental load, in 2003 he seriously posited the strategic idea of "the new industrialization road of the developing countries going through the "environmental mountain"," that is, digging a tunnel into the hillside of the "environmental mountain" to pass through, as shown in Fig. 3.1. In this manner, the mountain climbing activity has become a mountainous activity, while the paid price (environmental load) is low and the forward distance (economic growth) has not changed.

For a large number of developing countries such as China and India, this strategic thinking is an indicator to make changes immediately. If changes do not begin immediately, it may be too late when the decision is made in the future.

3.2.3 Is It Possible to "Cross the Environmental Mountain"? A Case Analysis

Is it possible to go through the hillside of the environmental mountain? To analyze this question, Dr. Mao Jinsu separately selected commercial energy consumption and GDP as the resource load and development level of a national economic system

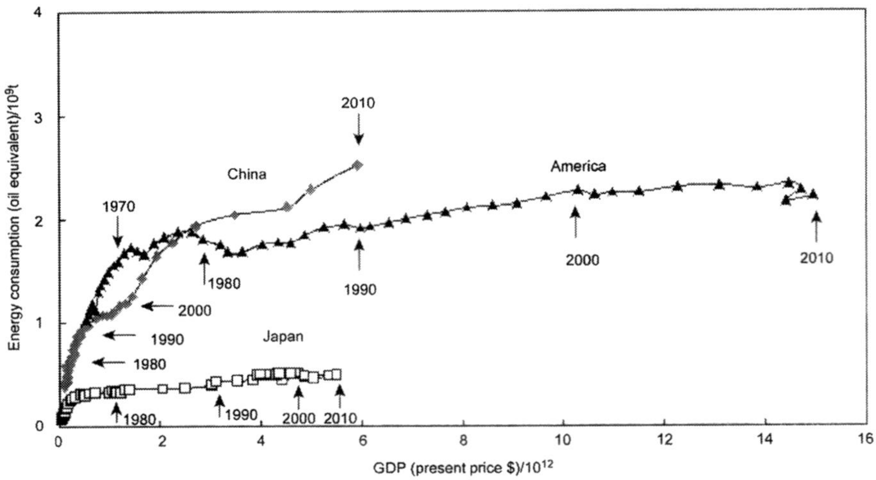

Fig. 3.2 Energy consumption of China, the USA, and Japan—GDP Curve. Reprinted from [5], with permission from Engineering Sciences (Chinese). Sun Mengying update

and studied the relationship between energy consumption and economic growth in Japan, the USA, and China between 1960 and 1996. The results are shown in Fig. 3.2. As seen from Fig. 3.3, the US and Japan's annual energy consumption had been basically stable in the two years at the end of the twentieth century. However, Chinese energy consumption and GDP grew together. Only at the end of the twentieth century did the rate of increase in energy consumption slow. If the data from the USA and China are connected into a curve, China continues to climb the "environmental mountain." Japan's situation is very different and its energy consumption has never risen to a high number. It can be argued that on the issue of energy consumption, the way that Japan moves forward is going through the "environmental mountain."

In short, the goal of environmental management is continually making progress in economy, culture, and technology. We should maintain the desired level of environmental load to achieve coordinated development between mankind and the environment.

3.3 What Is the Quantitative Relationship Between Human Beings and the Environment?

The third paragraph of this chapter mentioned the IPAT equation, which was earlier used to consider the quantitative relationship between human beings and the environment. With the advancement of sustainable development strategies, the method to measure the relationship between mankind and the environment has also

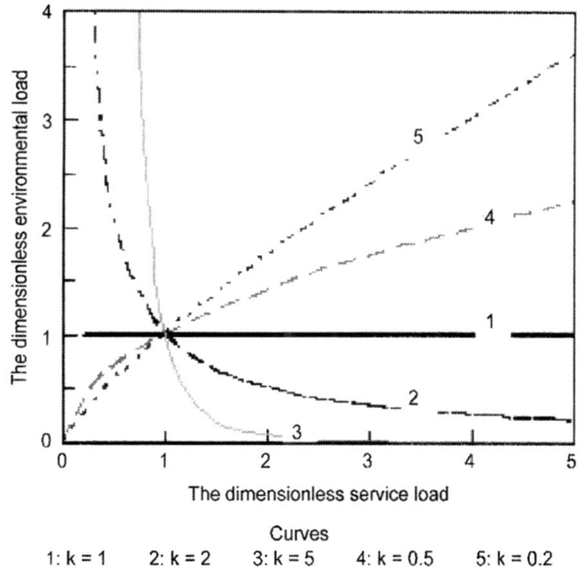

Fig. 3.3 Curve in which environmental load changes with social service. Reprinted from [8] with permission from Acta Scientiarum Naturalium Universitatis Pekinensis

been improved. At the turn of the century, the World Business Council for Sustainable Development (WBCSD) proposed the concept of "eco-efficiency" as an important indicator of the basic relationship between mankind and the environment, which was incorporated into environmental comprehensive management.

3.3.1 What Is Ecological Efficiency?

Ecological efficiency (or eco-efficiency) refers to the amount of social services that an industry or economic system can provide when it generates a unit of environmental load, reflecting the "technical" level that the system uses to achieve specific human services. If the eco-efficiency, environmental load, and social service volume of the system are represented by symbols *e*, *I*, and *S*, respectively; the definition of ecological efficiency can be expressed as follows:

$$e = \frac{S}{I} \quad (3.2)$$

where *I* and *S* are the environmental load and social services generated by the corresponding system. Social services can be expressed as the output of products, social GDP (or GNP), and other forms of the system in the specified statistical period. While the environmental load can be expressed as the consumption of resources, the discharge of environmental pollutants and other forms. Accordingly, based on selected different types of environmental load, the eco-efficiency can be expressed differently. For instance, when the environmental load is the resource consumption, ecological efficiency can be called resource efficiency and when the environmental load is environmental pollutant emissions, the ecological efficiency can be called environmental efficiency. Because of the different types of ecological efficiency, their physical meaning and units are also different.

It is easy to see that Eq. (3.2) can be reformatted as the following:

$$I = \frac{S}{e} \tag{3.3}$$

In contrast to Eqs. (3.1c) and (3.3), it is not difficult to determine that when the total social welfare in the IPAT equation is represented by social service S [the product of P and A in Eq. (3.1c)], T will reflect a system that obtains the environmental load generated by the unit service. From the definition point of view, this has precisely a reciprocal relationship with the "eco-efficiency" in the numerical value. Equation (3.3) can be seen as another expression of the control equation, which can be called an ISE equation in application.

In application, GDP is often selected as the social service of a country or region. At that time, the resource efficiency is transformed into the economic output corresponding to the unit consumption of the system and the environmental efficiency becomes the economic output corresponding to the unit environmental pollutant of the system.

It should be noted that the definition of eco-efficiency in Eq. (3.2), the IPAT Eqs. (3.1a–3.1c) and its variants (3.3) are for the same system.

3.3.2 Under What Circumstances Will the Environmental Quality Improve? A Theoretical Analysis

To effectively analyze the variable relationship between human development and environmental load, the ISE equation is expressed in a dimensionless form:

$$\overline{I} = \frac{\overline{S}}{\overline{e}} \tag{3.3a}$$

where $\overline{I}$, $\overline{S}$, and $\overline{e}$ are the dimensionless environmental load, social service, and ecological efficiency, respectively, which are equal to the ratio of the corresponding values in year t to the corresponding values in the base year. Equation (3.1c) reflects the relationships among environmental load, social GDP, and changes in ecological efficiency over a period of time.

Assuming that social service and ecological efficiency change exponentially with time, the following is true:

$$\overline{S} = e^{\rho_s \cdot t} \tag{3.4}$$

$$\overline{e} = e^{\rho_e \cdot t} \tag{3.5}$$

where ρ_s and ρ_e are the annual change rate of social service and ecological efficiency, t is an integer of not less than 1 from the base year, and other symbols are the same as previously described.

Solving Eqs. (3.3a), (3.4), and (3.5) simultaneously, they can be collated as follows:

$$\bar{I} = \bar{S}^{1-k} \tag{3.6}$$

where k is the ratio of the annual rate of change in ecological efficiency to the annual rate of change in social service, i.e., the rate of variable speed.

When social service and ecological efficiency grow exponentially, Eq. (3.6) represents the environmental load changes with the social service volume. To more clearly observe the relationship between environmental load changing and socio-economic development, Eq. (3.6) is depicted as a curve in which environmental load changes with social service, as shown in Fig. 3.3.

The following can be seen from Fig. 3.3:

(1) When k equals 1, the relationship between $\bar{I}$ and $\bar{S}$ can be represented as a horizontal line of 1, which reflects that when the ecological efficiency and economy grow simultaneously, the environmental load will not change with the growth of social services. When an area has good environmental quality and its environmental load is still within the environmental carrying capacity, its environmental quality will maintain a basically stable state, which can be called the environmentally stable area. This situation is more frequently occurring in developed countries or regions that have completed industrialization and urbanization. However, in areas where the environmental load has exceeded the environmental carrying capacity, some environmental impact will lead to the deterioration of environmental quality over time.

(2) When k is greater than 1, the relationship between $\bar{I}$ and $\bar{S}$ can be represented as a set of descending curves, as curves 2 and 3 show in Fig. 3.3. It shows that environmental load will decrease with the increase in social service volume when the growth rate of ecological efficiency is higher than the growth rate of social service. In this case, it is possible to reduce the environmental load to the environmental carrying capacity within its range; therefore, the environmental quality is improved. Moreover, it can be seen from the upper and lower positions of curves 2 and 3 that the larger the value of k, the faster the environmental load is reduced. It can be concluded that there is a possibility that environmental quality will gradually improve along with economic growth only when the rate of ecological efficiency is higher than the rate of economic growth. Moreover, the greater the gap is between these two growth rates, the greater the likelihood of environmental quality improvement. This scenario is the hope and goal of environmental management.

(3) When k is less than 1, the relationship between $\bar{I}$ and $\bar{S}$ can be represented as a set of ascending curves, as curves 4 and 5 show in Fig. 3.3, indicating that the

environmental load will increase with the increase in social service volume, which means that the environmental condition will be further deteriorated. Moreover, it can be seen from the upper and lower positions of curves 4 and 5 that the smaller the value of *k*, the faster the environmental load grows. It can be concluded that when the growth rate of ecological efficiency is less than the rate of economic growth, it will inevitably lead to deterioration of environmental quality with the increase in social service volume. Moreover, the greater the gap is between these two growth rates, the more obvious the deterioration of the environment. This situation is happening frequently in the rapidly industrialized and urbanized developing countries or regions.

It can be verified that if the social service and ecological efficiency are changed according to the linear law, the same qualitative conclusion can be obtained. However, in that case, the changes in each indicator will be slightly smaller.

In application, the change speed is higher than the value of k, which can be used as the criterion of the relationship between human socio-economic development and the environment. The greater the value is of *k*, the more conducive to environmental improvement.

In addition, it should be noted that the decrease in environmental load does not mean that the quality of the environment will be improved. Because the environmental load has been exceeded, even if the annual environmental load begins to decrease, if the environmental load is not reduced below the carrying capacity, the amount above the carrying capacity will continue to accumulate in the environment, causing environmental quality to continue to deteriorate. Only when the environmental load decreases to a point below the environmental carrying capacity will the environmental load accumulated in the original environment be released and relieved; however, this situation will continue for many years before the accumulation of all the loads disappears and the environmental quality is fundamentally improved.

3.4 How to Perform Environmental Management

During economic development, it has been observed that the performance of the environmental load first increases and then decreases in an inverted U-shaped with the economic growth, which provides evidence for environmental management and achieving the "win-win" of economic growth and environmental quality in the future. However, this is not the natural transformation of the economic system; it requires human beings to perform "careful planning, reasonable arrangements" and joint efforts with technology, consumption, management and other aspects. Therefore, understanding the variation law of environmental load and implementing scientific planning and management for environmental status in accordance with the law have become important elements in environmental management.

3.4.1 Environmental Load Characteristic Curve

(I) Expectation of the environmental load curve

Reduction of the environmental load in social and economic development cannot be achieved overnight but must be planned with a phased implementation. First, environmental management and technical measures should be employed to achieve a rise in environmental load and the separation of economic growth, called "de-coupling" or "uncoupling." Then, the growth rate of the environmental load must be made to be lower than the growth rate of service; thus, gradually reducing the environmental load to a desired level.

To simply design the curve of the environmental load from increase to decrease, assume that the service grows linearly and the annual growth rate is, so after the base year, the social service change multiplier in year t is as follows:

$$\overline{S} = 1 + \rho t \tag{3.7}$$

While the environmental load increases at a lower rate, the annual growth rate is $(1 - \varphi t)\rho$, in which the letter φ is the coefficient of change rate of the environmental load. Under such conditions, the environmental load varies:

$$\overline{I} = 1 + \rho t - \varphi \rho t^2 \tag{3.8}$$

It is easy to observe that Eq. (3.8) can be expressed as a parabola of the environmental load with time. In Fig. 3.5, ρ equals 0.1 and φ equals 0.02.

(II) Characteristics of environmental load points

As seen from Fig. 3.4, changes in the environmental load generally pass through the following four specific locations:

(1) Compared with economic growth, the environmental load began to increase at a slower rate, and the environmental load line began to "separate" from the economic growth line; therefore, point A is called the separation point. It is located in the starting year of environmental management.
(2) Point B is the highest point on the environmental load line and is known as the daily limit. At this point, $\frac{\partial \overline{I}}{\partial t} = 0$. Because the environmental load increases the year prior to point B and decreases year by year after point B, point B is the turning point of the environmental load.
(3) Point C is the critical point at which the environmental load returns to the base year after experiencing increases and decreases. Then, the environmental load will be lower than the environmental load of the base year, so it is called the drop point. At this point, $\overline{I_C} = \overline{I_A} = 1$.

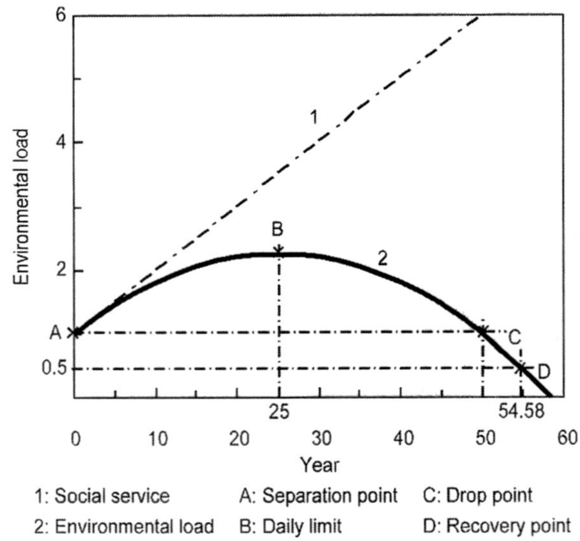

Fig. 3.4 Characteristic curve of environmental load over time. It is assumed that the environmental load in the base year is twice as high as the carrying capacity of the environment, social service is linearly increased at an annual rate of 0.1, and the variation coefficient of the environmental load is 0.02. *Source* Mao [9]

(4) Point D is the critical point at which the environmental load is reduced to the environmental carrying capacity during a long decline. After this point, because the environmental load is lower than the environmental carrying capacity, the excess environmental load that has originally accumulated in the environment will continue to be released and the environmental quality will continue to improve; therefore, this point is called the recovery point. If the environmental load in the base year exceeds the environmental carrying capacity *kI* times, the environmental load at point D is $\bar{I}_D = 1/k_I$.

The size of the environmental load at the four points (A, B, C, and D) is an important indicator of the environmental quality trends, which have an extremely important significance to environmental management. They are known as the environmental load characteristics points.

3.4.2 Curve of Ecological Efficiency

As mentioned previously, the improvement of environmental quality is not a natural transformation of the economic system. It can be seen from Eq. (3.3) that there must be a substantial increase in ecological efficiency.

To obtain the quantitative results of the ecological efficiency growth, Eqs. (3.7) and (3.8), which are the environmental load and the social service volume changing with time, can be substituted into the dimensionless ecological efficiency relation in Eq. (3.3a), resulting in the following:

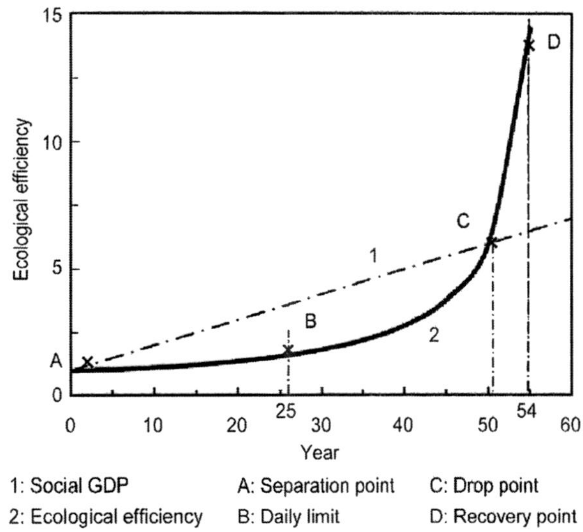

Fig. 3.5 Characteristic curve of ecological efficiency with time. It is assumed that the environmental load in the base year is twice as high as the carrying capacity of the environment, social service is linearly increased at an annual rate of 0.1 and the variation coefficient of the environmental load is 0.02. *Source* Mao [9]

$$\bar{e} = \frac{1 + \rho t}{1 + \rho t - \varphi \rho t^2} \tag{3.9}$$

If we assign specific values to φ and ρ, mapping Figure based on Eq. (3.9), then we will obtain a curve under the conditions of dimensionless change with time and a solid line as shown in Fig. 3.5. It is a curve that rises over time and increases in slope. During the graphing process, ρ equals 0.1 and φ equals 0.02. It can be seen that the process of decoupling and reducing environmental load and economic growth will result in sustained substantial growth in ecological efficiency.

3.4.3 Characteristic Curve of Environmental Management

(I) Relation curve of environmental load changing with social service

In environmental management applications, it is often necessary to understand the curve of the relationship in which the environmental load is changing with social service. Using Eqs. (3.7) and (3.8) in conjunction and eliminating time t, we obtain the following equation:

$$\bar{I} = \left(1 + \frac{2\varphi}{\rho}\right)\bar{S} - \frac{\varphi}{\rho}\bar{S}^2 - \frac{\varphi}{\rho} \tag{3.10}$$

If we continue to use values of 0.1 for ρ and 0.02 for φ and graph the figure based on Eq. (3.10), we obtain Fig. 3.6. The meaning of each feature point is consistent for Figs. 3.4 and 3.5.

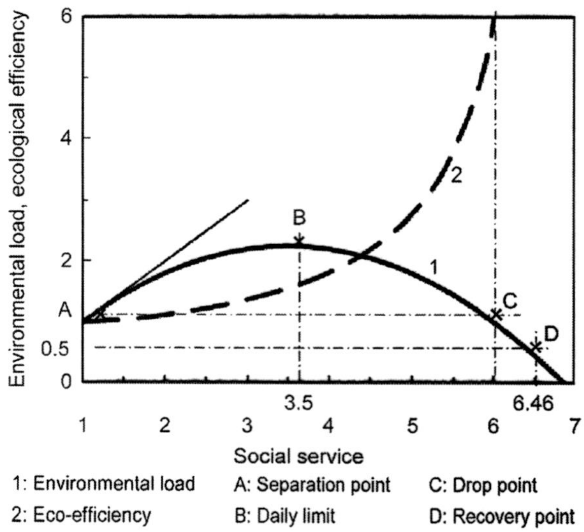

Fig. 3.6 Curve of environmental performance parameters changing with social GDP. It is assumed that the environmental load in the base year is twice as high as the carrying capacity of the environment, social service is linearly increased at an annual rate of 0.1, and the variation coefficient of the environmental load is 0.02. *Source* Mao [9]

The curves such as environmental load and ecological efficiency change with time, and the environmental performance parameters change with social service or economic development are all called as the environmental management curves.

(II) Characteristic Values of Environmental Management

The year of the environmental management feature point, social service volume, environmental load and ecological efficiency of the characteristics, and other values are the important node data in environmental planning and management, known as environmental management eigenvalues.

Based on the characteristics of the environmental load values at the feature points described previously, the year of occurrence of each feature point can be solved by Eq. (3.8). Substituting their values into Eqs. (3.7), (3.8), and (3.9), we obtain the corresponding multiples of social service, environmental load, and ecological efficiency. The calculation results are summarized in Table 3.1.

3.4.4 Example: The Management Plan for the Chinese Lead Environmental Load

The annual GDP growth rate in China was approximately 7% in 2000, and the number of industrial cycles of lead exceeded the natural circulation of lead 12.9 times. The goal is to develop a long-term plan for environmental management and

Table 3.1 Environmental management eigenvalues

Feature points	Separation point	Daily limit	Drop point	Recovery point
Code simple	*A*	*B*	*C*	*D*
Year of occurrence (*t*)	0	$\frac{1}{2\varphi}$	$\frac{1}{\varphi}$	$\frac{\rho \pm \sqrt{\rho^2 + 4\varphi\rho(1-\frac{1}{k_I})}}{2\varphi\rho}$
Multiple of social service	1	$1 + \frac{\rho}{2\varphi}$	$1 + \frac{\rho}{\varphi}$	$1 + \frac{\rho \pm \sqrt{\rho^2 + 4\varphi\rho(1-\frac{1}{k_I})}}{2\varphi}$
Multiple of environmental load	1	$1 + \frac{\rho}{4\varphi}$	1	$\frac{1}{k_I}$
Multiple of ecological efficiency	1	$1 + \frac{\rho}{4\varphi + \rho}$	$1 + \frac{\rho}{\varphi}$	$k_I\left[1 + \frac{\rho \pm \sqrt{\rho^2 + 4\varphi\rho(1-\frac{1}{k_I})}}{2\varphi}\right]$

Source Mao [9]

predict the changing conditions of the environmental load to reduce the environmental load of lead to be within the environmental carrying capacity within two generations (50 years).

Solution:

Based on the goal, $\rho = 0.07$, $\bar{I}_0 = \bar{I}_A = 1$, and $k_I = 12.9$.

Reducing the environmental load of lead to be within the environmental carrying capacity within 50 years:

$$\bar{I}_{50} = \frac{1}{12.9} = 0.078$$

Substituting the value of $\bar{I}_{50}$ into Eq. (3.8) results in the following:

$$0.078 = 1 + 0.07 \times 50 - \varphi \times 0.07 \times 50^2$$

The solution is $\varphi = 0.025$.

Substituting the values of ρ, φ, and k_I into Table 3.1 to obtain the environmental management eigenvalues provides the results listed in Table 3.2.

It can be observed that to reduce the environmental load of lead to be within the environmental carrying capacity within 50 years, it is necessary to increase the ecological efficiency by 58.05 times in the same period.

To develop goals for every five years, substitute the values of and k_I into Eqs. (3.7), (3.8) and (3.9), and calculate the values of environmental performance parameters for each year. The calculation results are listed in Table 3.3.

It should be mentioned that the status quo of the ecological efficiency of lead in China and Sweden has been analyzed in Ref. [6] from the perspective of lead efficiency in lead-acid batteries. The results show that the eco-efficiency of lead in the Swedish lead-acid battery system is more than 100 times the ecological efficiency of lead in China. This huge disparity shows that it is possible to achieve a 58× improvement of the ecological efficiency of lead in China within 50 years.

Table 3.2 Characteristics of lead environmental management in China

Feature points	Separation point	Daily limit	Drop point	Recovery point
Code symbol	*A*	*B*	*C*	*D*
Year of occurrence (*t*)	0	20	40	50
Multiple of GDP	1	2.4	3	4.5
Multiple of social service	1	1.7	1	0.0775
Multiple of ecological efficiency	1	1.41	3	58.05

Table 3.3 China's lead environmental performance parameters over 50 years

Planning year	2000	2005	2010	2015	2020	2025	2030	2035	2040	2045	2050
GDP multiple	1	1.35	1.7	2.05	2.4	2.75	3.1	3.45	3.8	4.15	4.5
Environmental load multiple	1	1.31	1.52	1.65	1.69	1.64	1.51	1.28	0.97	0.57	0.08
Ecological efficiency multiple	1	1.03	1.12	1.24	1.42	1.67	2.06	2.69	3.92	7.31	58.01

3.5 Method to Promote Human Sustainable Development

Sustainable development involves human activities and resources and the environment, and all aspects of interaction between the two, so that sustainable development can be pursued from a variety of perspectives. Following is an outline of the history of human beings' effect on the material.

From the perspective of human activities, it is necessary to adjust the human consumption concept, improve technology, abandon poor production and lifestyle, improve the mode of economic operation, and strengthen the system construction, fundamentally changing the relationship between human activities and the environment.

3.5.1 Resource Protection and Rational Use

Natural resources are the basis for human survival. Based on the physical types of resources, natural resources include land, climate, water, biological, mineral, marine, energy, and tourism resources. Based on the scarcity of resources, natural resources include depleted and non-exhaustive resources. Among them, non-exhaustive resources are also known as "renewable natural resources," including constant resources (such as solar energy) and resources that are easy to misuse and pollute. Depleted resources include renewable and non-renewable resources, primarily including land, regional water, and biological resources.

The characteristics of natural resources are limited, regional, holistic, and versatile. Therefore, the implementation of sustainable development must provide effective protection and rational use of natural resources to maintain sustainable use, in particular, effectively protecting non-renewable and scarce resources. The principle of resource protection is that the consumption rate of natural resources should not exceed the speed of natural resources or the speed of the development of alternative resources so that natural resources can be used in the production and life of current and future generations.

Different protection measures should be taken for different natural resources. For example, for mineral resources, we should not only strengthen geological prospecting work to obtain additional available resources but also fully understand their limits and actively enhance the technical level and improve the efficiency of the comprehensive utilization of resources. For scarce resources, we should actively seek alternative resources for their protection. For ecological resources, we should apply different measures to protect them and their service functions and health status. For instance, we should remove or mitigate human disturbance as soon as possible and simultaneously perform actions to restore original structures and functions of ecosystems that have undergone serious damage. In the case of ecosystems that have been destroyed and cannot be restored to their original structure and function, artificial ecological design should be adopted to assist in ecological alteration or reconstruction. For an ecologically sensitive system with unique landscape and special ecological service functions, we should establish ecological protection zones, effectively avoiding brutal interference and damage.

3.5.2 Sustainable Consumption

The goal of human activities is to meet the needs of mankind. The process of human consumption is to meet the needs of production and life. According to different consumers, consumption can be divided into production-based and practical consumption. Production-based consumption is usually manifested as the process of material transformation; thus, it can be included in the production process. The consumption referred to in this section is meant to satisfy the process of human end-user needs, including the enjoyment of products or services.

In the final analysis, satisfying the growing needs of mankind is not only the source of human development momentum but also the source of stimulating production needs, the services of resources and environmental protection. It is not difficult to imagine that different consumption patterns of mankind and different consumption will affect resources and environmental systems. For example, choosing to live in small houses or live in large houses will affect the use of the resource of land and resources used for construction; therefore, we can promote the implementation of sustainable development strategies with the help of sustainable consumption patterns.

Sustainable consumption refers to the provision of services and related products to meet the basic needs of mankind and improve quality of life, while minimizing the use of natural resources and toxic materials, thereby generating the minimum amount of waste and pollutants in the service or product life cycle so as to not damage the needs of future generations. Sustainable consumption should follow the following principles: moderate consumption, fair consumption, and people-oriented consumption. The moderate consumption principle refers to the level of human consumption that should begin with the necessary needs of mankind, reducing excessive consumption that is beyond human needs and maintaining human health, thereby reducing human consumption of natural resources and destruction of the ecological environment. The fair consumption principle states that the consumption level of all people should be controlled within a certain range, avoiding a small number of people with high consumption, ahead of consumption, squandering consumption, and abnormal consumption. The people-oriented consumption principle refers to the formation of a reasonable consumption structure, focusing on the spiritual level during the process of meeting peoples' normal material consumption needs, and avoiding a small number of people in developed areas that focus excessive attention on material consumption that not only wastes resources but also causes a reduction in a spiritual civilization. These principles can be reflected in daily life, such as increasing the temperature of air conditioning as high as possible in the summer and taking public transport when traveling as much as possible.

3.5.3 Circular Economy

Circular economy is the economic form that has material circulation as its core in the human social and economic system. It imitates the natural circulation of material during the entire process of resource investment, enterprise production, and product consumption, and it has the ability to organize and optimize man-made flows of material as well as to achieve economic savings by conserving natural resources and reducing waste.

Following the "reduction, reuse, resource-based" principle, a recycling economy relates organically to the human socio-economic system, such as human consumption and industrial production, resource use, and environmental pollution. In the use of resources, a circular economy emphasizes the efficient use of resources, comprehensive utilization of resources, and waste recycling of resources to achieve a reduction in natural resource consumption, thus saving and protecting natural resources. During the production process, the circular economy not only emphasizes "less material, more efficient" but also promotes product life by developing a durable and maintainable product. It also considers recycling production waste as well as recycling the changes of waste among enterprises. Waste that cannot be directly used will be recycled. For consumer segments, the recycling economy encourages consumers to choose environmentally friendly products and emphasizes the return of products to the producers or marketers to help waste become resources after the product is scrapped. Finally, for waste that can no longer be reused, emphasis is placed on the use of an end of process to ensure that the waste is returned to the environment in an environmentally safe manner. See Chaps. 4–11 for details.

3.5.4 Environmental Governance

Human activities inevitably discharge various waste materials into the environment. Because of the transformation process of these substances through human activities, when a resource returns to the environment, its state of existence is significantly different from its original state as a resource found in the environment. Some waste can be converted into natural systems through environmental migration, while some waste requires manual treatment to assist in effectively integrating the waste into the natural system. Additionally, if the emissions of this waste are higher than those of the natural system, some waste will accumulate in the environment, resulting in the increase of certain substances in the environment, and the deterioration of environmental quality. Therefore, we must use environmental management tools to manage and improve.

Environmental management in sustainable development is reflected in the following. The waste generated in the production process should be recycled and used as raw materials for other production processes, thereby reducing the amount of environmental waste. Conversely, for waste that cannot be recycled, the waste can be made harmless with the help of end treatment technology, such as conventional dust removal, catalytic conversion, incineration, placement in landfills, curing and other measures, such as generating heavy metals geological polymers, which are being researched and developed. In addition, for the ecological environment that has been polluted and destroyed, such as soil, we can use ecological measures including plant absorption, inhibitors, redox control, and different tillage methods to perform scientific environmental management and restore the original environmental quality level.

3.5.5 Improvement of the Political System

The political system is the basic guarantee for safeguarding the security and interests of human society. Promoting sustainable development must be included in the improvement of the political system.

In 1979, China enacted the Environmental Protection law of the PRC (for trial implementation) and revised it in 2014. It clearly stipulates environmental supervision and management, protection and improvement, and pollution prevention and control and provides a legal guarantee for protecting and improving the environment, preventing pollution, safeguarding public health, and promoting ecological civilization construction and sustainable economic and social development. The production link is the primary link over which human activities affect resources and the environmental system. To improve the situation of resources and the environment, the Standing Committee of the National People's Congress passed the Cleaner Production Promotion Law in June 2002. That year, increasing the status of resources and environmental protection from a management perspective, China formally established the Ministry of Environmental Protection and strengthened the overall coordination of environmental protection, comprehensive management, publicity and education, and other aspects of work. China promulgated the People's Republic of China Circular Economy Law, which states that the circular economy refers to reduction, reuse, and resource-based activities during production, circulation, and consumption. It provides the basic management system, incentive measures, and legal responsibility in the implementation of the circular economy and establishes the basic contents of reduction, reuse, and resource utilization. Additionally, its legal status increased the importance of the circular economy and promoted sustainable development.

The Ministry of Environmental Protection has also provided a series of ISO 14000 standards for environmental management in the management of technology, including an environmental management system, environmental audit, environmental labeling, and life cycle analysis. The goal is to guide the various organizations to achieve the proper environmental behavior. It not only can strengthen the government's level of environmental management but can also improve the competitiveness of enterprises in its product market and promote export trade. Additional details are provided in a later chapter.

Classroom Discussion and Assignments

Issues for Discussion

In the Chap. 2, the groups discussed and analyzed the status of environmental problems of interest. This discussion will examine the objectives of environmental management for the next 30 years from the perspective of environmental management based on the results of the previous assignment, thus providing a basis for understanding the various approaches to sustainable development strategies in subsequent chapters. Therefore, the topics discussed in this chapter are as follows:

Based on the environmental issues selected in the previous chapter, we develop environmental management plans for the next 30 years for a particular area.

Considering that students may choose from several types of environmental problems, such as resources, environmental pollution, and ecological damage, the following are proposed recommendations from the themes in this chapter:

Resource management planning;
Ecological management planning; and
Environmental management planning.

Example of the Process of Screening Issues

Based on the results of the screening of the previous environmental issues combined with the requirements of the previous chapter on the selected topics, the selected topics should be limited as much as possible, such as the study of soil remediation planning or atmospheric environment planning in a specific area.

Using the study of atmospheric environmental planning as an example, the screening process is shown in Fig. 3.7.

Possible Research Programs and Content

First, select a typical location, such as Beijing, based on the air quality report.

Current research includes environmental capacity, pollution status, excess multiple, source of pollution, emission intensity, and other research. Based on the human health atmosphere environmental standards, the environmental capacity is allowed to have a PM2.5 environmental capacity at the case site. Compare the actual concentration and the environmental capacity. The environmental capacity can determine the multiple of the PM2.5 pollutants in the status quo. The current status of the pollution can be obtained through environmental monitoring, including existing public monitoring methods, such as the Environmental Protection Agency or the Meteorological Bureau or researcher distribution, personal monitoring, and other means to obtain the pollutant concentration data. Pollutant sources and emission intensity can be obtained by means of on-site exploration, pollution sources classification and analysis and measurement.

The plan is used to establish as a goal the number of years required to correct the haze, i.e., reduce the haze to levels below the environmental capacity within 30 years.

Planning phase analysis can be divided into a planning period of every 5 or 10 years, analyzing the environmental performance at various stages, including the changing relationship of the environmental load, ecological efficiency, and social service and the influencing factors of these indicators, as well as the changing relationship of the influencing factors required to achieve the goal.

The guarantees that are needed by planning, management, and recommendations and by analyzing the overall objectives include laws, regulations, and policies and improvements in technology and facilities, the ecological environment and social and economic changes.

Important sources of information related to environmental quality are recommended below.

http://www.zhb.gov.cn/

China Environmental Protection Department: water quality, atmospheric environmental quality, soil environment, ecological environment and other information, http://www.zhb.gov.cn/.

IAQ http://www.cma.gov.cn/

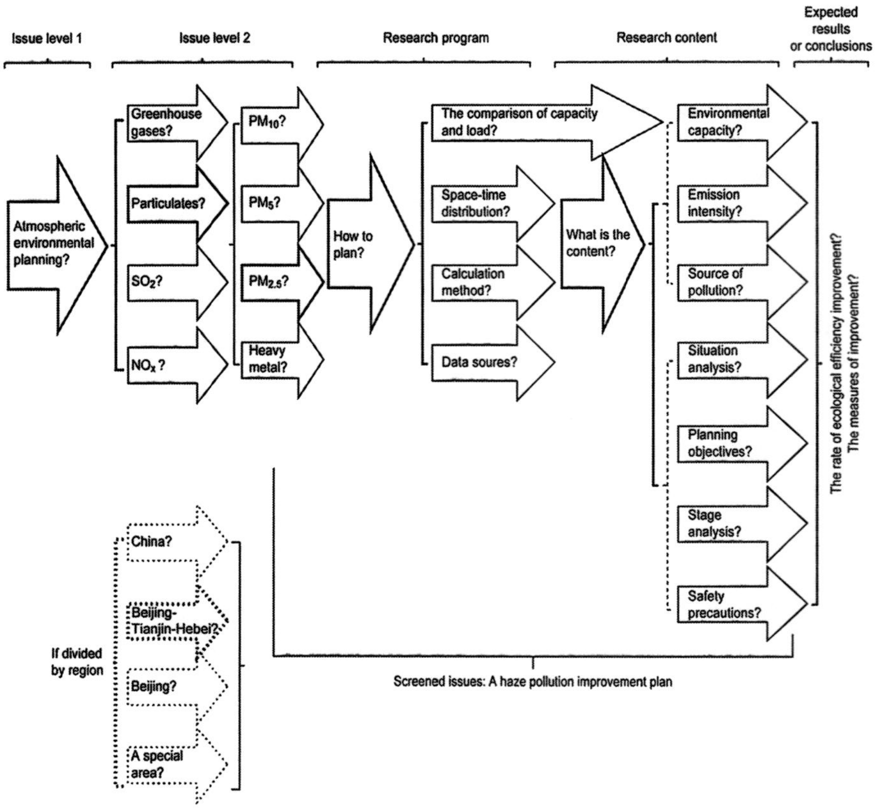

Fig. 3.7 Screening process for the discussion topic and the analytical framework of the analytical process

China Meteorological Administration: including precipitation, air pollution weather conditions (such as the IAQ index, air pollution diffusion conditions) and other information, http://www.cma.gov.cn/.

http://data.mlr.gov.cn/

China Ministry of Land and Resources: containing mineral resources, land resources, marine resources and other bulletin information, http//data.mlr.gov.cn/.

http://www.sac.gov.in.

China National Standardization Management Committee can check the national standards, international, and other national standards (including other various professional standards established by the Ministry of Industry and Information Technology, Ministry of Environmental Protection and the Health and Family Planning Commission, http://www.sac.gov.in).

Requirement of Staged Scientific Research

This stage requires students to learn to focus on discussion topics and to establish a logical relationship between topics and research programs and their content and to understand why the study can be conducted. What would you plan to study? How would you perform these studies? What are your initial thoughts concerning the type of expected results and conclusions? Focus should be on the validity of the data used in the study and the rationality of the findings of the study. The results of the study are presented in the form of PowerPoint reports, which should be concise and presented within 10–12 min. Before mid-semester, emphasis is placed on developing logical thinking and clearly expressing ideas. After mid-semester, emphasis is placed on the comprehensive use of knowledge points, innovative thinking and the normalization of expression.

Recommended and Referenced Literatures

1. Chinese Academy of Sciences Sustainable Development Research Group. 2014. *2014 report of China sustainable development strategy*. Beijing: Science Press.
2. Xuedong, Gong. 2001. *Environmental management*. Beijing: China Environmental Science Press.
3. Lin Aiwen, Hu Jiangjun, Zhang Ling, et al. 2005. *Resource environment and sustainable development*. Wuhan: Wuhan University Press.
4. Graedel T.E., and B.R. Allenby. 2003. *Industrial ecology*, 2nd ed. New Jersey: Prentice Hall.
5. Zhongwu, Lu, and Mao Jinsu. 2003. Crossing "Environmental Mountain"—On the increase and decrease of environment impact in the process of economic growth. *Engineering Science* 5 (12): 36–42.
6. Wenhu, Ye. 2001. *Introduction to sustainable development*. Beijing: Higher Education Press.
7. Publications by Prof. Dr. Friedrich Schmidt-Bleek. http://www.factor10-institute.org/publications.html.
8. Mao, J.S., et al. 2007. The relationship between industrial development and environmental impacts in China. *Acta Scientiarum Naturalium Universitatis Pekinensis (Natural Science)* 43 (6): 744–751.
9. Jiansu, Mao. 2003. *Industrial metabolism of lead and its impact on national economy*. Shenyang: Northeastern University.

4 Cyclic Flow of Material: Concepts and Classifications

4.1 Presentation of Core Issues

Resource shortages and environmental deterioration are primarily attributed to the disturbance caused by human activities beyond the natural capacity, and the primary purpose of human activity is to satisfy the material and cultural demands of people, which means that the primary object of human activity is "material." By using material resource processing, transformation, transportation, and other social and economic activities, the service goal of human demands can be satisfied. Therefore, improvement of resources and the environment should first focus on the "material" flow. What is the flow of material? What are the basic laws of material flow? How can we assess the flow state of material? How can we optimize and manage material? These issues are the theoretical basis for the circular economy development and will be discussed separately in the next three chapters.

This chapter focuses on the topic of material flow and poses the following questions: What is material flow? How does material flow occur? How does the flow affect the circular economy? What is its relationship with the resource environment to expound the concept, classification, and significance of material circulation?

4.2 What Is the Natural Flow of Material?

4.2.1 The Concept of the Natural Flow of Material

The universe is composed of material that is constantly moving. The movement of a substance into a different storage space according to a specific law is known as the flow of material. It is called a cycle of material when it begins at one storage space and returns to the same space after having been in several other spaces. The circulation between different storage spaces is called the cyclic flow of material.

J. Mao et al., *Circular Economy and Sustainable Development Enterprises*,
https://doi.org/10.1007/978-981-10-8524-6_4

Physical movement without any interference from human beings and their activities is called the natural flow of material, and the flow that is disturbed by human activities, especially to meet human demands, is called the anthropogenic flow.

For a better understanding of the natural flow of material, the concept of natural circulation of material in several related disciplines is reviewed. In ecology, people are concerned about the material relationship between the individual, community, ecosystem, and external environment; therefore, the natural circulation of material refers to the cycle of transfer and transformation of material between the biological and abiotic environments in the biosphere. In geochemistry, people focus on changes in composition, i.e., the morphological structure of the material in the earth, and the natural circulation of material primarily refers to the cyclic operation between the earth's different spheres such as atmosphere, lithosphere, hydrosphere, biosphere, and pedosphere. Another example of natural circulation is the process by which nutrients obtained from the atmosphere, water, and soil environments are absorbed by green plants and are part of the ecosystem for use by other creatures and then return to environment through decomposition in the biogeochemical cycle. It can be seen that different subjects have different research objects and service goals and the definitions of material natural circulation are different; however, all material flows are caused by natural forces.

4.2.2 Classification of the Natural Cycle of Material

There are many methods, influenced by subject and research goal, to classify the natural circulation of material. In this course, we focus on materials that have important significance for human survival, life, and production, such as fossil fuels as a type of mineral resource and metal as a type of base material, and divide them into the following three types of cycles: water, nutrients, and poisonous and harmful substances, based on the effects of chemicals on human beings. The following are examples of several typical materials in ecology from literature [1] to assist students in understanding the natural cycle of material.

(I) Water cycle

Water is an important component of life, and obtaining and maintaining stable and adequate water from the water cycle is necessary for human and ecosystem survival.

In the water cycle, the primary reservoir of material is the ocean. The sea water evaporates under the influence of solar energy and enters the atmosphere and then returns to the ocean and earth in the form of atmospheric precipitation. The precipitation landing on the earth follows several paths, including becoming soil water or groundwater after seeping into the ground; returning to the ocean through underground runoff; falling onto plants and returning to the atmosphere through transpiration after being absorbed by the plants; evaporating directly into the atmosphere from the ground; and returning to the ocean through surface runoff. The water that returns to the ocean evaporates into the atmosphere from its surface, forming the natural cycle of water.

In addition, the water cycle performs the following functions for human beings: delivery of nutrients, regulation of climate, and maintenance of ecosystem stability. Within the water cycle, many nutrients can be transported into the ecosystem including various organs of the body because nutrients can be dissolved in water, making it crucial in providing nutrients to the ecological system. The water cycle can regulate the climate because the specific heat of water at normal temperature is 4.18 kJ/(kg °C), which means that approximately 4180 kJ of heat will be released per ton of water under the unit temperature difference. Water is commonly used in engineering as a heat carrier for the heat transfer process, such as an indoor heating system or a cooling tower. Additionally, the evaporation heat of water under normal pressure is approximately 6300 kJ/kg, meaning that 6.3 GJ of heat will be released during the evaporation of 1 t of water. Therefore, the natural circulation of water is important in regulating temperature, increasing humidity, and improving climatic conditions. Water is the primary component of the ecosystem, and its natural cycle is an important guarantee of ecosystem stability, thereby supporting and ensuring the health and stability of the human life system.

(II) Nutrients—the natural cycle of carbon

Carbon is the nutrient element of living systems. The circulation of carbon has been a significant global research topic in recent years because of the environment problem of global warming.

The carbon cycle in the ecosystem begins with the photosynthesis of the producer, through which the carbon dioxide in the atmosphere is stored in organisms and passes to other organisms through the food chain. In this process, one portion of the carbon translates into biological components and the biological excreta or carbon in biological residues translate into carbon dioxide by decomposition and are released to the atmosphere or converted to fossil fuel with the release of carbon dioxide through burning. The remaining carbon is released into the atmosphere through the respiration of various plants and animals, forming a cycle of carbon from the atmosphere and back to the atmosphere. Concurrently, carbon dioxide exchanges at the interface of the atmosphere and hydrosphere by diffusion, so that the carbon content in the ecosystem can be self-regulated and repaired, maintaining a balance. According to statistics, approximately 10^{11} t of carbon dioxide is absorbed by the water and the same amount of carbon dioxide will return to the atmosphere from the water every year.

(III) Toxic and hazardous substances—lead circulation

When lead pollutants, which are hazardous chemicals, reach specific concentrations in the environment, they can cause harm to people, other organisms, and the environment, and have long been a concern of environmental geosciences and other fields.

The lithosphere in the earth's crust contains the largest repository of lead. The natural circulation of lead begins with rock weathering and volcanic eruption and is lost in the atmosphere as dust particles. This accounts for approximately 85% of the total lead in the atmosphere and is the main source of natural emissions of lead into the atmosphere. Other natural emissions are a result of plant exudates (approximately 10%) and forest fires (approximately 3%). In the atmosphere, lead and its compounds exist primarily in dust and aerosol states, and they will enter a body of water or return to the ground with rainfall, dust, and other processes. In water, lead primarily exists in the forms of $PbOH^+$, Pb^{2+}, and $PbSO_4$, and it easily reacts with CO_3^{2-} and SO_4^{2-} to form insoluble compounds such as $PbCO_3$ and $PbSO_4$. The concentration of lead is generally not more than 20 mg/L in contaminated water. It is primarily transported in water as suspended solids. With the flow of water, the concentration of lead will rapidly drop to a few micrograms per liter or a few tenths of a microgram, and much of it is deposited into the mud. The soluble lead content in soil is very low and exists primarily in the forms of $Pb(OH)_2$ and $PbCO_3$ that will be deposited slowly. Plants can absorb the lead ions Pb^{2+} in the soil and accumulate more in their roots. Then, the migration and circulation of lead in the atmosphere, water, soil, and organisms can be completed, which is called the biogeochemical cycle of lead or the natural circulation of lead. As shown in Fig. 4.1, the natural circulation rate of lead is usually expressed as approximately 28 thousand t/a by atmospheric emissions in its natural state.

4.2.3 Basic Characteristics of the Natural Flow of Material

The natural flow of material has the following characteristics:

(I) A circular movement is performed in the specific ecosystem or among the natural spheres of material. With the material transfer along the biological trophic level in the ecosystem, the waste produced by an organism is the food of another organism. A complex network structure is established with the interaction of material flow in different organisms forming a closed circular flow of material in the ecological system.

(II) The natural force is the impetus for the natural flow of material in the natural system. Although various space migrations and morphological transformations will occur in the natural flow of material, the power of these changes come from natural forces, including changes internal to the material, such as the chemical relationship among molecules and atoms when combining with other materials in the external environment and the concentration difference, pressure difference, gravity or heat that may be experienced during the diffusion and sedimentation of materials. These changes occur in natural places, such as the atmosphere and bodies of water.

(III) Material flows among different pools at a specific flux rate. A location where the amount of material is clearly greater than that of other places is called a

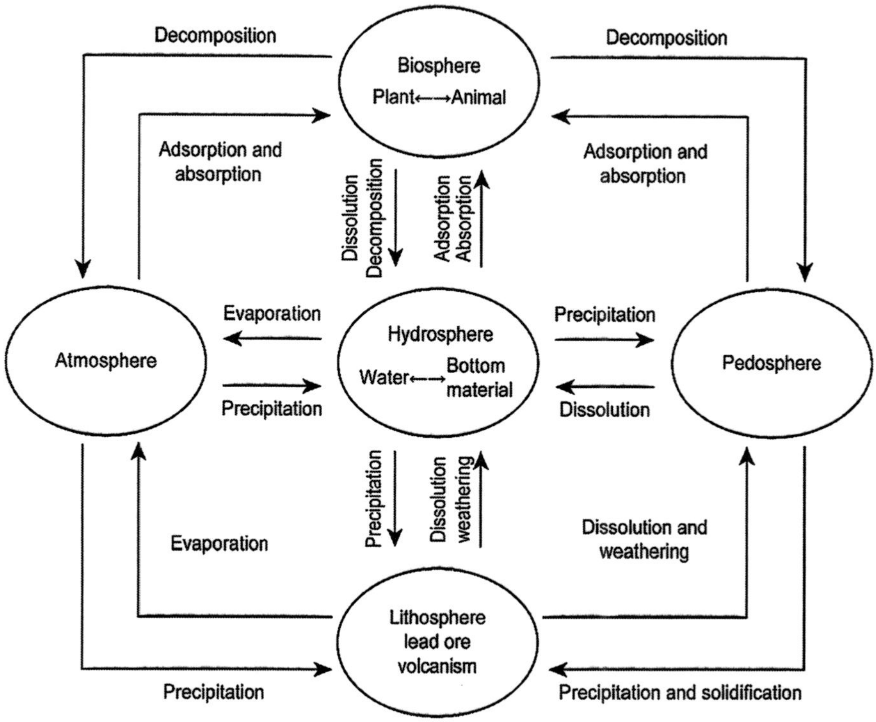

Fig. 4.1 Geochemical cycle of lead. *Source* Qi et al. [2]

reservoir; for example, the hydrosphere is a reservoir for water and the lithosphere is a reservoir for many metals. The natural flow of material is primarily presented as the flow among reservoirs, and the circulation quantity of material per unit of time is called the flux rate, which is used to measure the exchange level among the reservoirs. On the premise of constant material inventory, the greater the flux rate, the faster the material turnover. However, for reservoirs that continue to decrease with the flow of material, the faster the flux rate, the faster the inventory is depleted.

4.2.4 Ecological Significance of the Natural Circulation of Material

The natural cycle of material is the basis of ecosystem metabolism and is of great significance to maintaining continued stability of the ecosystem in the following manners:

(I) Transport of essential elements needed for the ecosystem—Material is the basis of the ecosystem, and the development of the ecosystem is inseparable from the basic substances that constitute life. For example, the water cycle can provide freshwater sources for terrestrial organisms, freshwater organisms, and human beings. Additionally, many nutrients are water soluble, complete their migration with the water cycle, and are utilized by organisms.

(II) Maintenance of the distribution stability of material in all ecosystems—The natural circulation of material has a specific rate, thus ensuring that material distribution in different ecological subsystems is maintained at a specific rate, with neither a lack nor accumulation of certain substances in an organism, maintaining the materials that are essential for life at a reasonable level and a certain stability of the material structure in each system.

(III) Maintenance of stable ecosystem functions—The material needed for an ecosystem is delivered based on the required quantity and quality through the natural circulation of material, and the waste from ecosystem metabolism, especially some toxic and harmful materials, is transferred to the environment in a timely manner. If waste is not promptly transferred out of the ecosystem, there may be excessive accumulation in the body, which will show symptoms of poisoning when the concentration exceeds the biological endurance capacity, therefore threatening the ecosystem's health. The natural circulation of material can effectively guarantee the ordering and structural stability of material in the ecosystem and stabilize the ecosystem function.

(IV) Maintenance of human survival and sustainable development—Adequate material storage capacity and a suitable consumption rate are required to maintain the resource demand for human survival and sustainable development, and the effective recycling of material is important to remove pollutants and maintain all types of material components in the environment system at the appropriate levels. For example, with adequate water resources and humid air in Beijing, particulate in the atmosphere is likely to be absorbed into the water, fall to the ground, and finally flow away with surface runoff, which would improve the quality of the atmospheric environment. However, atmospheric particles are not easily absorbed under dry conditions and remain in the atmosphere, resulting in deterioration of atmospheric quality.

In general, the natural circulation of material is the basis of human and earth existence and is also an important mechanism for sustainable development. It primarily involves subjects such as ecology, geography, and geology as well as related derived secondary disciplines and branches such as geochemistry and environmental chemistry. See Recommended References at the end of this chapter.

4.3 What Is the Anthropogenic Flow of Material?

4.3.1 Concept of the Anthropogenic Flow of Material

As mentioned previously, the anthropogenic flow of material is related to the natural circular flow, which is the process of material flow arising from human production and life activities to meet demands. It is called an anthropogenic circular flow when a material returns to its original source through a number of other socio-economic processes from the initial human activity.

Material that is important for human production and life will be the focus of the circular economy because the anthropogenic flow of material is generated to meet human demands. Next, we use metallic material as an example of anthropogenic circulation.

First, the lithosphere is a metal repository. The metallic material must be removed from the metal mineral resources to meet the demands of human production and life and enter the human socio-economic system from its natural circulation to begin the anthropogenic flow, as shown in Fig. 4.2. Second, after mining, smelting, and other productive processes, the metal can be transformed into metal materials, such as steel and aluminum. After further processing, metal products with specific functions will be formed, such as automobiles and bridges. Then, the products enter the social systems after market transactions to fulfill the products' service functions. The product service life ends after a period of time, and the product is scrapped as waste. Part of the product forms waste resources through recycling and returns to the production process for regeneration, and the remainder is discharged into the natural environment in the form of waste and pollutants, thus completing the anthropogenic flow of material by returning to the natural ecosystem

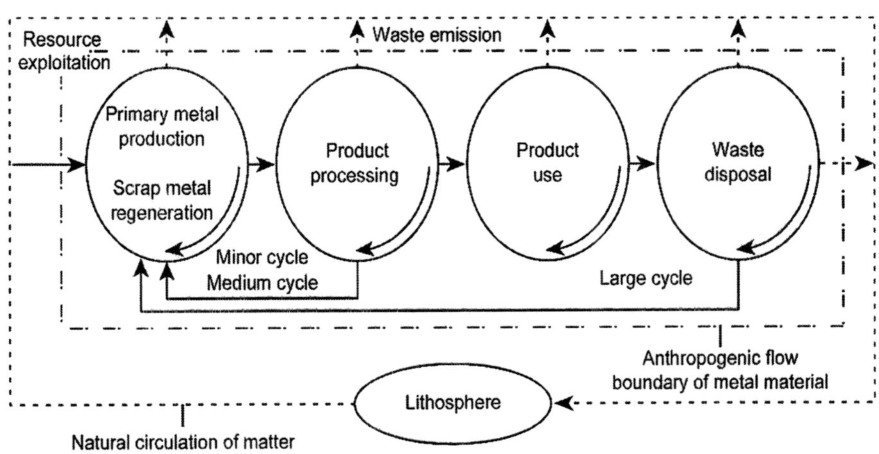

Fig. 4.2 Conceptual framework for the relationship between anthropogenic circulation and the natural circulation of metallic materials. Reprinted from Mao et al. [3]

and entering the natural circulation of the material. In the anthropogenic recycling of material, it is not possible to make every molecule in every production link into products due to technology limitations, thus forming waste and pollutants. Part of the waste and pollutants will enter the natural environment, and the remainder is waste resources that are recycled and returned to production for regeneration, forming the human circular flow of material.

4.3.2 Classification of the Anthropogenic Flow of Material

The anthropogenic flow of material cannot only be classified based on the categories of materials but also on the manner of material circulation.

It can be divided into the anthropogenic circulation of simple and compound materials based on the material structure. Simple material refers to most metals (such as iron, copper, aluminum, and lead) and ecosystems nutrients (such as nitrogen and phosphorus). Compound material can be divided into simple compounds (such as H_2O), complex compounds (such as the majority of synthetic engineering materials, including plastic and paper), and synthetic chemicals (such as PCBs and organochlorine).

Based on the availability of materials to humans, it can be divided into resource flow ("resource exploitation" shown in Fig. 4.2 as the flow of resources into the anthropogenic system), waste flow ("waste emission" shown in Fig. 4.2 as the flow that discharges to the environment from the anthropogenic system), and the flow in each stage of the product life. Additionally, it can be divided into circular and non-circular flow of material based on whether the material can return to its original life cycle stage. The anthropogenic circular flow of material can be divided into three categories of large, medium, and minor cycles based on the different productions that flow through them, and the amount each sector is used in the human socio-economic systems or the length of their life cycles as shown in Fig. 4.2.

(I) Minor cycle—This refers to the material circulation within an enterprise, such as the waste return from a downstream process to upstream processes for recycling as a raw material; for example, if a large steel plate is needed to make the baseboard of a ship, it results in large pieces of scrap during processing that may be reused in other workshops because it is not sufficient to produce another baseboard or side but is sufficient for a wicket of the side plate or the floor of a canoe. This cycle of material within an enterprise is generally called a minor cycle, includes the recycling of water, other consumables, and by-products, and is only performed by manufacturing enterprises.

(II) Medium cycle—A medium cycle is a material cycle among enterprises, such as waste being used as a raw material that returns from a downstream industry to an upstream industry or the use of industrial waste or surplus energy that is sent to other industries. Medium cycle refers to the process of reducing waste produced during recovery processing being returned to the material production department for cyclic regeneration to form new industrial materials. For example, the remaining scrap from the minor cycle within manufacturing enterprises is collected and returned to the production department for material regeneration. Usually, the medium cycle involves many production enterprises and their waste recovery departments.

(III) Large cycle—Some of the material is returned to the production department for reuse as raw material after the product is used and scrapped. Using metal as an example, the smelting process for metal material can be considered as the beginning of an anthropogenic circular flow. Metal is formed and becomes products with specific functions after processing and manufacturing for subsequent purchase and use by the consumers, such as a building, a computer, or a cup. The product is scrapped and recycled, used for other purposes, or returned to the waste metal recycling sector for reproduction of metal material after several years of use, forming the cycle from metal manufacturing to production of metal known as the large cycle. In general, large cycle not only involves a variety of manufacturing enterprises and waste recovery departments, but also involves social consumption.

The three different cycles involve and are managed by different departments. Extended producer responsibility (EPR) is advocated in a circular economy, which means the responsibility extends to the producers in the recovery stage after the material is scrapped. Specifically, the producer attaches a label to new products to ensure maintenance is performed during use and the products are recycled by the enterprise after the consumers' use. As another option, a special company (a material recycling company) is established to collect the local waste and old materials that are uniformly classified and returned to the corresponding production department. The minor cycle is naturally performed within the enterprise; therefore, management should be performed in different regions based on the characteristics of the cycle.

In addition, different cycle types occur over different time durations. A minor cycle is usually several days, the medium cycle may be several months, and the large cycle could be years or even centuries. Because the majority of products can have a service life of a few years, decades, or even hundreds of years (i.e., ten years for ordinary household appliances and 70 years for architectural design), coupled with the time for material production, product manufacturing, logistics trade, and transportation, the duration may be a few years or even hundreds of years for a large cycle.

In application, the anthropogenic circulation of specific material and regions at different levels is studied based on the type of material and the purpose, duration, and location of the study.

4.3.3 Basic Characteristics of an Anthropogenic Flow

The anthropogenic flow of material has the following characteristics:

(I) The impetus for anthropogenic flow is human demand in the human socio-economic system. The satisfaction of human needs is the original impetus for an anthropogenic flow. To satisfy these needs, a series of space migration and morphological changes in the human socio-economic system, the results of mutual action between human activity and material, must be completed by the technologies of design, production, and transportation, all of which occur in the human socio-economic system.

(II) The anthropogenic circular flow of material is a special form of anthropogenic flow and is the result of optimized management for environment improvement. The basic process of anthropogenic flow is the movement from natural resources to product (or service) and, finally, to waste. The amount of resource that fails to turn into a product will become industrial waste material, which will return to the natural systems through environmental emissions. Secondary resources are treated as waste instead of natural resources within a circular flow, and the relationship between waste and resources is closed, which not only forms the material circulation but also achieves cost savings and emission reduction.

(III) The anthropogenic flow rate and material cycle rate are important indexes to reflect the anthropogenic flow of material. The anthropogenic flow rate refers to the amount of anthropogenic flow in a unit of time, including the natural resources consumption and the waste discharged into the environment. The amount of a certain material resource exploited or the amount of emission of environmental pollution in a year reflect the disturbance intensity of human activities on the natural system. The sustainable utilization of resources or the change of environmental quality can be analyzed by comparing these indexes with natural carrying capacity or natural cycle rate. The recycling rate shows the proportion of circulation in an anthropogenic flow of material and can be estimated as the ratio of the secondary resources that return to resource regeneration to the total inputs, which reflects the resource saving condition or waste utilization level. There are other ways to express this rate; see Chap. 5 for additional details.

(IV) In-use stock can be formed by the anthropogenic flow of material and is a location for the storage of used material. The total amount of material in the in-use stock is called the material usage inventory. With increasing human demands, the amount of material in the in-use stock increases when it is greater than the amount used during the same statistical period. This will lead to an accumulation of material if the situation continues. The amount of material in the in-use stock is the material quantity that migrates into the natural system by human activities and is the result of human interference with the natural system. The per capita accumulated usage also reflects regional social consumption patterns and material consumption levels. See Chap. 6 for additional details.

4.4 What Is the Relationship Between the Anthropogenic Flow of Material and Natural Circulation?

The fundamental difference between the anthropogenic flow of material that is caused by human activities and the natural circulation of material that is caused by nature is the impetus for the material flow. The material circulation of an anthropogenic flow is the foundation of circular economy development, but it is restricted by the human management level, technical level, and natural circulation of material, which reflect the complex and interactive relationships among human society, the economic system, and nature.

Comparing the anthropogenic flow and the natural circulation of material, it is not difficult to see that the anthropogenic flow of material can be seen as an important link in the natural circulation of material. However, it is the interference of human activities on the natural system that results in the demand for natural resources from the natural environment, with the resulting large emission of waste and pollutants into the environment as shown in Fig. 4.3. The different effects on material caused by different human production processes lead to a divergence of the human activity circle and the natural sphere and the redistribution of material in the earth's surface system. It is not difficult to understand that material in natural resources and waste and pollutants are distinct in form, quality, concentration, and location. Therefore, the anthropogenic flow of material generates a negative impact on the natural surface system, especially when the anthropogenic flow rate is beyond the natural circulation rate of material. This will not only interfere with the natural circulation of the material but also cause a continued accumulation of the material that natural circulation cannot absorb into the environment. Certain material in the environment will threaten the health of the ecosystem when its concentration exceeds the endurance capacity of certain creatures in the biological ecosystem. Therefore, by analyzing the anthropogenic flow of material, we cannot

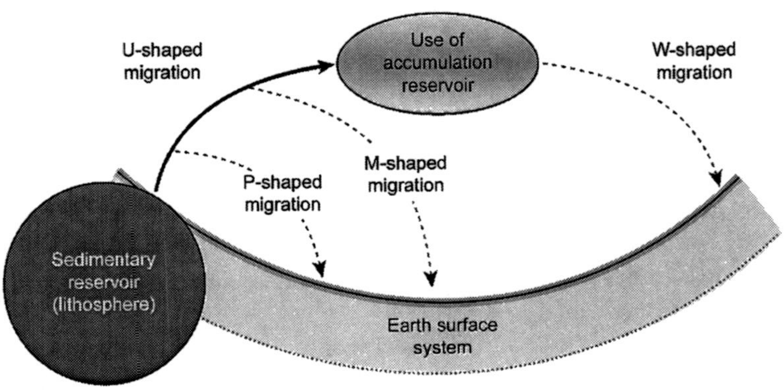

Fig. 4.3 Conceptual framework for the relationship between the anthropogenic and natural circulations of metals. Reprinted from Mao et al. [4]

only identify the disturbances and their intensity on the natural system from human activities, which provides an important basis for the management of saving resources and reducing environmental pollution emissions, but we can also understand how human socio-economic activity interferes with material status. This provides insight into methods to improve human activities through scientific management and realize the harmonious development of human beings and the environment. Therefore, it is of great significance for effective promotion of sustainable development.

Classroom Discussions and Assignments

Topic for Discussion

The environmental issues of interest were screened by groups in the previous two chapters, and the majorities are closely related to a specific substance circulation; thus, the following are the topics of this chapter: Which material circulation is the most worthy of attention? How are natural and anthropogenic flows conducted? A qualitative description is required.

Example of Topic Selection

Topics can be selected based on the type of material, such as phosphorus or nitrogen in nutrients or heavy metal substances such as lead or cadmium. They can also be selected based on the representative pollutants exposed in the typical regional pollution problems, such as carbon related to global warming and nitrogen related to eutrophication of a lake.

The research protocol can be proposed by each student and shared by learning the research methods in the published thesis. An example of the screening process is shown in Fig. 4.4.

Thesis Title and Introduction Example

Example of a Thesis Title

The following are the requirements for a thesis title defined in the Chinese Core Journals of *Environmental Science*:

The title should be concise and accurately reflect the content, generally not more than 20 words, see http://www.hjkx.ac.cn/hjkx/. Examples of recommended thesis titles are as follows:

Metal elements content, pollution, and geochemical characteristics of the sediments in the Songhua River.

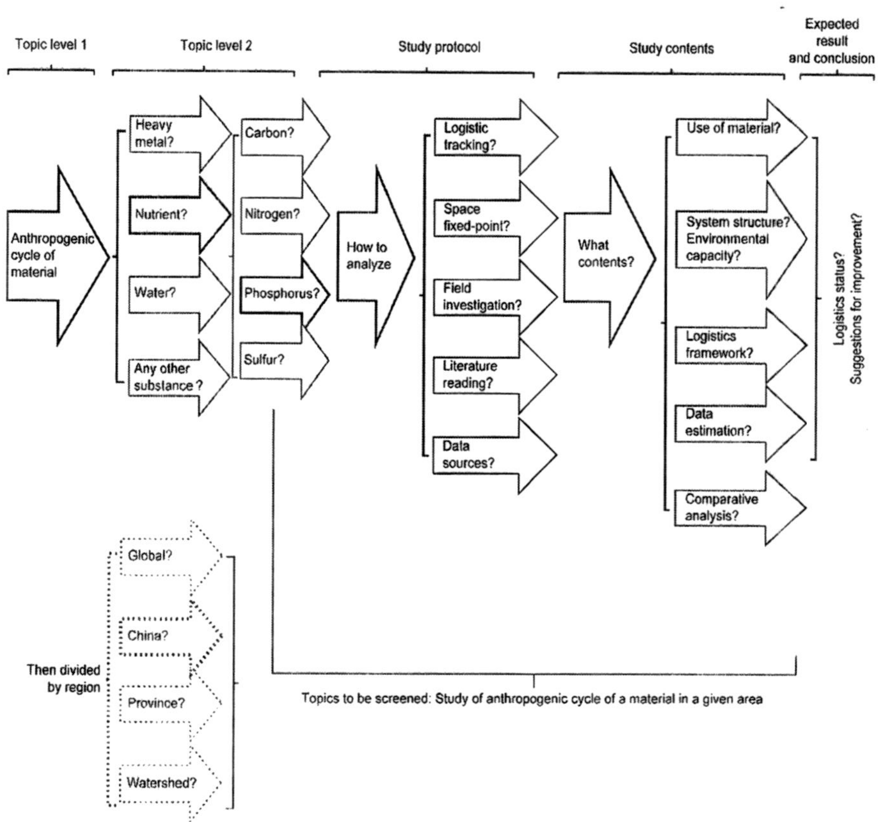

Fig. 4.4 Selection of class discussion issues and process analysis framework

Effect simulation of the wetland water changes on the calamagrostis angustifolia litter decomposition and nitrogen dynamics.

Study of suspension, sedimentation, and endogenous release of sediment under the action of wind and waves.

The topic should be clearly stated in the thesis along with the research methods and contents.

Thesis Introduction Example

The requirements for an introduction (also called a preface) to a thesis defined in Environmental Science are used for reference. The preface includes the related work of domestic and foreign predecessors (use of a quotation is acceptable) and the purpose, characteristics, and significance of this work. Do not repeat the scientific knowledge.

(I) Example 1

The following is the Introduction in the thesis "Crossing the 'environmental mountain'-: discussion of the increase and decrease of the environmental load during economic growth."

Since the founding of new China, especially the reform and opening to the outside world, China's economy has continued to grow at a rapid pace and currently has a considerable economic aggregate. Concurrently, the environmental loads have become quite significant, resulting in a grim environmental situation.

In recent years, a series of vital measures in environmental protection have been adopted in China, achieving remarkable results. "The national environmental situation is transformed from the overall deterioration of the environment quality and improvement in part to the basic control of the intensified trend for environmental pollution and improvement of the environment quality in some cities and regions."

However, because China is still in the process of industrialization and it will be a significant amount of time before the entire process is completed, the method for China to achieve the win-win of economy and environment is the major issue that must be decided now. This is because of the following: (a) The rapid economic growth is likely to continue for many years, and only proper action can avoid the rapid rise of the environmental load; otherwise, China's environmental problems may be very serious in a few years. (b) China is the largest developing country, with a specific percentage of the world's total economic and environmental load, which will significantly increase in the future. If the total amount of environmental load cannot be effectively controlled, not only will it affect our country, but it will have a significant impact on the world.

The 16th National Congress of the Communist Party of China indicates that a new industrialization path with good economic returns, low resource consumption, less environmental pollution, and full use of human resources must be initiated.

This contains foresight, provides a grand strategic goal that indicates the forward path for the people of the nation, and is of great significance. The task of environmental protection and economic workers is to provide further detail and execute the new industrialization path.

The view that environment and development must be linked together to identify the essence of the problem has been gradually realized since the 1980s. Recently, a number of monographs have been published both at home and abroad that are of great reference value. For example, the results of model calculations are applied to the literature to provide scenario predictions of world population, resources, and environmental change over the next few decades.

Based on existing literature, this article will discuss the principles of avoiding serious environmental problems in the latter half of the industrialization of our country; perform a necessary theoretical analysis on the increase and decrease of the environmental load during economic growth; use energy consumption as an example to analyze the changes in environmental load during economic growth in some countries and provinces in our country; and forecast the future environmental load of our country.

From: Lu and Mao [5].

(II) Example 2: an introduction example for a topic that includes new terms

For relatively new research that may involve new terms or vocabulary, the topic must first be interpreted to provide clear meaning of the new terms. For example, the first paragraph in the thesis "Lead in-use stock: a dynamic analysis" follows.

Lead in-use stock refers to the lead currently providing services of various kinds to humanity. It results when more material enters use over time than leaves use. The amount of lead in-use stock for a region or country reflects that regions standard of living under currently available technology, therefore, lead in-use stock is an indicator of the amount of metal that less developed regions may need to put in place to attain comparable material services. Additionally, knowledge of the magnitude of lead in-use stock is useful in estimations of the amount of lead that will be available for future recycling.

From: Mao and Graedel [6].

The introduction should clearly explain why the research should be performed, the current research progress and its significance at home and abroad.

Recommended and Referenced Literatures

1. Li, Bo. 2003. *Ecology*. Beijing: Higher Education Press.
2. Liu, Qi, Huake Wang, and Wenda Li. 2005. The ecological effect and global chemical recycle of the heavy metal lead. *Journal of Anhui Institute of Education* 23 (6): 97–100.
3. Mao, Jiansu, Jing Liang, and M.A. Lan. 2014. Changes in the functions, species and locations of lead during its anthropogenic flows to provide services. *Transactions of Nonferrous Metals Society of China* 24 (1): 233–242.
4. Mao, J.S., M.A. Lan, and J.P. Niu. 2012. Anthropogenic transfer & transformation of heavy metals in anthrosphere concepts, connotations and contents. *International Journal of Earth Sciences and Engineering* 5 (5): 1129–1137.
5. Lu, Zhongwu, and Jiansu Mao. 2003. Crossing the 'environmental mountain'—discussion of the increase and decrease of the environmental load during economic growth. *China Engineering Science* 5 (12): 36–42.
6. Mao, J.S., and T.E. Graedel. 2009. Lead in-use stock: a dynamic analysis. *Journal of Industrial Ecology* 13 (1): 112–126.
7. Zhou, Qixing, and Guohong Huang. 2001. *Environmental biogeochemistry and global environmental change*. Beijing: Science Press.
8. Chen, Jingsheng, Baoshan Deng, Shu Tao, et al. 1990. *Environmental geochemistry*. Beijing: Ocean Press.
9. Yang, Zhifeng, and Jingling Liu. 2010. *Introduction to Environmental Science*, 2nd version. Beijing: Higher Education Press.
10. Mao, J.S., J. Dong, and T.E. Graedel. 2008. The multilevel cycle of anthropogenic lead I: methodology. *Resource Conservation and Recycling* 52: 1058–1064.
11. Mao, J.S., J. Dong, and T.E. Graedel. 2008. The multilevel cycle of anthropogenic lead II: results and discussion. *Resource Conservation and Recycling* 52: 1050–1057.

Circulation of Substances: Analysis of Logistics

5

5.1 Presentation of Core Issues

To clarify the concept of mankind's flow of material, this chapter will explore how material flows in the human socio-economic system and how the flow may be managed and optimized.

Through the discussion in the previous chapters, we find that there are different environmental problems in different regions, and different environmental problems involve different types of substances. The natural circulation process of different material types and the process of human flow are also different. Beginning with the simplest material, scientific research is often performed with the principle of "simple and complex"; therefore, this chapter selects a simple substance. Based on their relationship with modern social and economic development and the resources and environment of this type of material, we select metal substances as the representative material for this topic.

To scientifically answer this question, we require scientific research methods. Currently, there are two types of international material circulation flow analysis methods. One analysis method focuses on the material particle and tracking the entire human flow process, which is similar to the Lagrangian method of tracking fluid particles in fluid mechanics, referred to as "logistics tracking" or "tracking." This method was created by the Chinese academician Lu Zhongwu. The other method focuses on a region and analyzes the man-made flow process for the material of the region, which is similar to the Euler method in fluid mechanics for the fixed study of changes of the fluid physical parameters in a space point, referred to as "logistics fixed-point method" or "fixed-point method." This method was created by T. E. Graedel of Yale University. This chapter uses the tracking method as an example, successively introduces the analysis methods for material flow, and applies the law of material flow based on this method and management law.

J. Mao et al., *Circular Economy and Sustainable Development Enterprises*,
https://doi.org/10.1007/978-981-10-8524-6_5

5.2 Method to Analyze the Man-Made Flow of Material

5.2.1 What Is Material Flow Analysis?

Material flow analyses [substance flow analysis (SFA) and material flow analysis (MFA)] are focused on a substance and analyzing its numbers and directions during various segments (or life cycle stages) of the human socio-economic process, including the entire product life cycle from the mining of the natural resources to the end-of-life disposal of the products (or services).

Performing material flow analysis will focus on substances in the human socio-economic system and explore which links they experience. What is the sequence? How is the quantity of material distributed among the various links and stocks? What is the direction of the flow? By tracking the entire man-made flow of material process, we will clearly understand the various links, the sequences, and the flow directions that the material has experienced. However, to determine the quantity of each substance, we must follow the Material Conservation Law for each link, establishing the inflows and outflows of material and the material balance equation between the amount of material increase and decrease in the internal link and calculating the flow of non-available data based on available data. The available data are mainly taken from industry statistics or collected through field investigation. The calculation of some aspects requires use of physical and chemical relations.

5.2.2 Which Steps Should Be Followed in the Material Flow Analysis?

This section used lead as an example to demonstrate the basic steps of material flow analysis.

5.2.2.1 Establishing the Research Objectives and Selecting the System

The study of man-made material flow is to establish the quantitative relationship of the material human services between resources and the environment; additionally, it is to determine how to optimize the flow of material to reduce the impact of human activities on the environment. Therefore, when performing material flow research, we will first consider how to quantify the quantity of material, resource consumption, and environmental emissions.

When we previously introduced the quantitative relationship between human beings and the environment, we used resource consumption and environmental pollutant emissions in the statistical period to characterize the impact of human activities on external resources and the environment. Human services can be expressed as economic output, such as GDP, or product output or the number of final services provided by the product. However, the economic output often does not directly show the human need; for example, the service demand of "from A to

B" refers to specific conveyance that will safely transport people to a designated destination. The passenger consumption is the usability of the conveyance that can transfer the material "from A to B." Although there is a certain connection between the product and the service provided by the conveyance, it is a distinctive concept. A product is usually easier to quantify, while a product service is often more difficult to quantify. Therefore, the majority of material flow analyses use the product as the characteristic index of the service quantity provided by the system.

Although lead is widely used in mechanical, electronics, chemical, and other modern industrial areas, its primary use (approximately 80%) is in the production of lead-acid batteries. The service provided by a battery is transferring electricity, which can be quantitatively calculated, so it can be used to characterize the quantity of the battery services. Additionally, it is reported that the existing global lead from non-renewable resources (i.e., lead ore) can only be guaranteed to for the next 20 years, while China only has a guarantee of 10 years; therefore, there is a significant shortage in lead resources. Furthermore, the leaded waste and pollutants produced in the production process and after the lead products are scrapped may also threaten the health of the ecosystem, especially the health of the human body. Studies have shown that the human circulation rate of lead caused by human disturbance has exceeded its natural circulation rate by 12.9 times. Lead poisoning occurs in many areas, seriously threatening the health of the human body. These all indicate the selection of lead as a typical substance as well as establishing the quantitative relationship between human service and resources and the environment. Therefore, we select lead as a typical substance, use the lead-acid battery system as the object of our study of the man-made flow of material, and attempt to reveal the inherent quantitative relationship between human needs and resources and the environment.

5.2.2.2 Developing Assumptions and Building the Model

A lead-acid battery system includes several basic processes of lead production, lead-acid battery manufacturing, lead-acid battery use and battery scrap and recycling, known as the life cycle stages of a lead-acid battery. Lead moves through these stages, which are of varying duration. Following the principle of identifying the main problem, we assume the following. ① The duration of the production process is not considered. ② The average service life of lead-acid batteries is unchanged in the time duration of the study. ③ All the lead-acid batteries are scrapped after manufacturing year $\Delta\tau$. A portion of them are the depreciation of waste lead and return to the lead industry system for regeneration treatment in the year they are retired. ④ The study does not consider inventory problems in a certain stage. In this case, if the output of lead-acid batteries in year τ is P_τ, a flow diagram reflecting the flow direction and amount of lead for the year is drawn for the lead in the lead-acid battery based on the time sequence of the lead that flows through the stages, its quantity (which complies with the Material Conservation Law), and the inflow of lead that is equal to the outflow in each life cycle stage, as shown in Fig. 5.1.

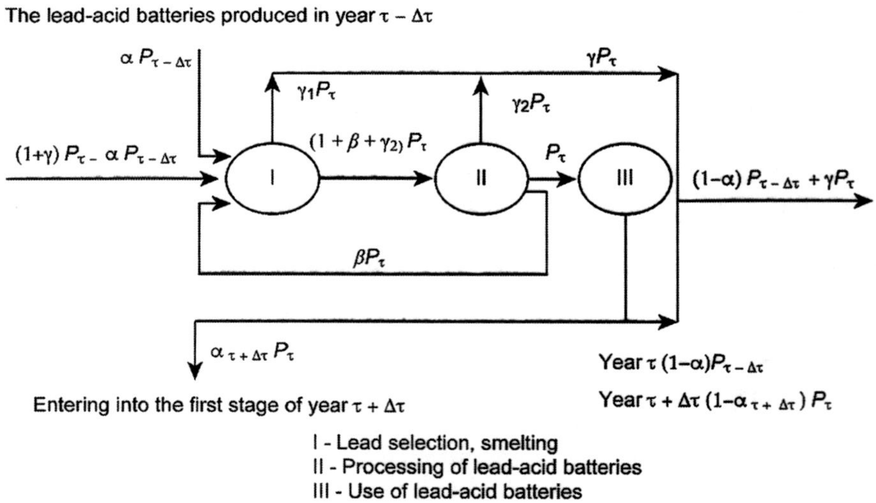

Fig. 5.1 Lead flow diagram of lead-acid battery life cycle. Reprinted from (Mao et al. [1]), with permission from John Wiley and Sons

The following should be noted in Fig. 5.1:

(1) Because the production of primary lead and the regeneration of scrap lead belong to the lead production process, we combine them. They are represented as symbol I.
(2) Setting the annual output of lead-acid batteries into constant change, the lead-acid battery production is $P_{\tau-\Delta\tau}$ in year $\tau - \Delta\tau$.
(3) The service life of a lead-acid battery is $\Delta\tau$ years, and the lead-acid batteries produced in year τ will be scrapped in $\tau + \Delta\tau$ years. However, the waste lead-acid batteries that are placed into operation in year τ are from the production of lead-acid batteries in year $\tau - \Delta\tau$.
(4) In the definition of lead-acid battery production, the proportion of lead returned to the lead in the refining phase after scrap is the large circulation rate of lead, which is expressed as the symbol α and its unit is t/t. In this case, there will be $\alpha_{\tau+\Delta\tau}P_\tau$ of lead returned to the refining phase of the lead-acid battery system in year $\tau + \Delta\tau$. During year τ, $\alpha_\tau P_{\tau-\Delta\tau}$ of the lead returns to the refining stage. For the sake of simplicity, the subscript symbol τ of the large circulation rate is omitted from the amount of waste lead placed into year τ in the figure.
(5) The definition of the circulation rate of lead is the amount of lead collected from the processed chip waste in the lead-acid battery manufacturing stage after considering the number of lead-acid batteries produced in the year, which is represented by the symbol β with the unit t/t.
(6) The definitions of the ratio of the number of lead emissions in the first and second life cycle stages and the number of lead-acid batteries produced during the corresponding life cycle are represented by symbols γ_1 and γ_2, respectively;

their units are t/t. For ease of use, the sum of the two emission rates is defined as the total lead emission rate of the production phase, referred to as the lead emission rate, expressed as symbol γ, where $\gamma = \gamma_1 + \gamma_2$.

5.2.2.3 Selecting the Parameters and Quantifying the Characterization

The environmental load and social service are selected to be the performance indicators of the environmental impact and human services, respectively. The limitations of the environmental load and social services are as follows.

Environmental load refers specifically to the impact of lead-acid battery systems on resources and environmental systems, divided into resource load and emission load. The resource load is defined as the number of lead ore resources consumed by the lead-acid battery system each year, denoted by the symbol R, and the emission load is defined as the number of lead-containing waste and pollutants discharged into the environment per year by the lead-acid battery system, expressed as the symbol Q. Both environmental loads are calculated in terms of lead content, and their unit is t. As shown in Fig. 5.1, the lead mining resource load and lead load in year τ are as follows:

$$R = (1 + \gamma)P_\tau - \alpha P_{\tau-\Delta\tau} \quad (5.1)$$

$$Q = \gamma P_\tau + (1 - \alpha)P_{\tau-\Delta\tau} \quad (5.2)$$

Because the main function of the lead-acid battery is to provide electricity to the electrical field, the total amount of electrical energy that can be provided in the service life of a lead-acid battery produced in a given year serves as the service of the corresponding year, expressed as the symbol S, with units of (kW• h) • a. It is not difficult to understand that lead-acid battery social service should not only be proportional to the service life of lead-acid batteries but should also be proportional to the lead content of lead-acid batteries, which can be expressed as follows:

$$S = F_P \cdot P_\tau \cdot \Delta\tau \quad (5.3)$$

where P_τ is the annual production of lead-acid batteries, calculated based on lead content, t; $\Delta\tau$ is the service life of lead-acid battery, a; and F_P is the ratio performance of a lead-acid battery, which is the scale factor between S and $P_\tau \cdot \Delta\tau$.

This method tracks specific products and considers the time between production and scrap to analyze the relationship between products, resources, and the environment, considering the time between production and scrap a significant feature, as proposed by academician Lu Zhongwu in 2000. This method has been used many times in research of the steel industry scrap resource problem and the source index of iron emission. Mao Jiansu transferred the focus from products to services provided by the products; thus, a quantitative relationship between human social service and the resource environment was established, which is an important method to perform material flow analysis.

5.2.3 Method to Apply Material Flow Analysis

Case analysis is used to demonstrate the application of material flow analysis. It can also be seen as the fourth step of material flow analysis, which is used to verify the feasibility of the above method and obtain the results. The lead-acid battery system continues to be used as the example.

The research background is the Chinese lead-acid battery system in 1999. China's domestic consumption of refined lead was 525,000 t that year, of which 66.8% (equivalent to 350.70 kt) was used in the production of lead-acid batteries. Based on the level of production technology at that time, in the production of lead-acid batteries, 0.92 t of lead was used, with an average input of 1 t of lead, 0.0356 t of lead was recycled as processing waste lead, and the remaining 0.0444 t of lead was released into the environment in the form of lead waste and contaminants. The average life expectancy of lead-acid batteries is estimated to be three years.

For waste lead recovery and regeneration, it is estimated that 90,900 tons of waste lead-acid batteries and 12,500 tons of processing waste lead were put into lead recycling, according to data gathered by the *China Nonferrous Metals Regeneration Association* in 1999. The lead recycling rate is estimated to be 80–88%, and we calculate the rate to be 86.37%. A total amount of 89,300 tons of renewable lead can be obtained, and the remaining 14,100 tons of lead are dissipated into the environment in the form of lead waste and pollutants. The remaining amount used for the production of lead-acid batteries is 2,614,000 t, based on the original amount of lead.

In the production of primary lead involving mineral processing, smelting, and other processes, the beneficiation yield was 83.8% and the smelting comprehensive yield was 92.78% in 1999, according to China's nonferrous metals industry statistics. Therefore, to obtain 261,400 tons of primary lead, lead ore that contains 336,200 tons of lead was needed, while 54,500 tons and 20,300 tons of lead were lost in the mineral and smelting processes, respectively.

In addition, because the average life expectancy of lead-acid batteries is estimated to be three years, the waste lead-acid batteries recovered in 1999 were produced in 1996. Table 5.1 lists the lead-acid battery production in China between 1990 and 2000. Considering Table 5.1 lead-acid battery production accounted for 75–85% of the total output, if the production in 1996 and 1999 was 77% and 78%, respectively, the production of lead-acid batteries can be estimated to have used 291,700 t of lead in 1996. Since the recovery of waste lead-acid batteries was 90,900 t in 1999, 200.77 kt waste lead-acid batteries failed to be recovered or were not included in the statistical data.

Table 5.1 China's refined lead and lead-acid battery production between 1990 and 2000

Year	1990	1991	1992	1993	1994	1995	1996	1997	1998	1999	2000
Lead-acid battery production/GWh	6.980	5.146	6.837	7.773	–	7.080	9.487	–	–	10.394	11.881

Source China Machinery Industry Yearbook

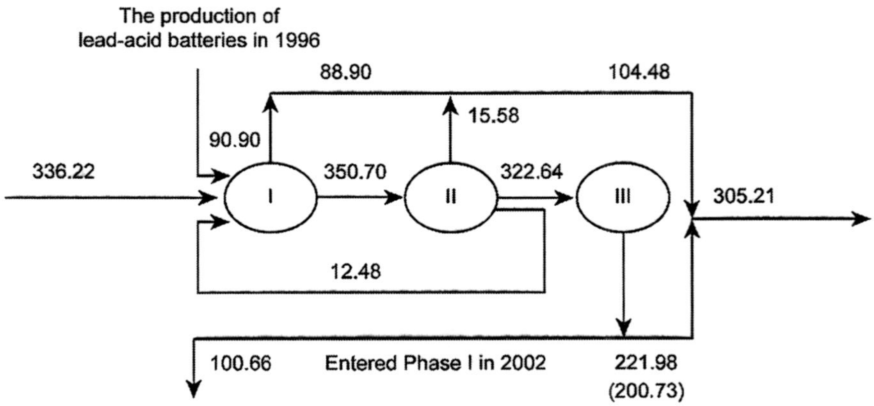

Fig. 5.2 Lead flowchart of China's lead-acid battery life cycle in 1999. Reprinted from (Mao et al. [1]), with permission from John Wiley and Sons

Based on the above data, China's flowchart of the lead-acid battery life cycle in 1999 can be arranged as shown in Fig. 5.2. The unit is kt.

The quantity allocation and flow direction of the stock flow in the lead-acid battery production system are shown in Fig. 5.2.

5.3 What Is the Law of Human Circulation Flow?

Chapter 3 introduced the concept of eco-efficiency, which refers to the number of social services that can be provided by a unit of environmental load in a product system, characterizing the relationship between human services and resources and the environment. The lead-acid battery system continues to be the example, and eco-efficiency is chosen as the index to reflect the relationship between the human element and the lead environment, to deduce the basic law of the lead flow in the lead-acid battery system.

As mentioned previously, the system has two types of environmental load, including lead mineral resource consumption and lead pollution discharge. Correspondingly, ecological efficiency can be divided into resource efficiency and environmental efficiency, which are expressed by r and q, with the unit of (kw·h)·a/t:

$$r = \frac{S}{R} \tag{5.4}$$

$$q = \frac{S}{Q} \tag{5.5}$$

It can be observed that the higher the ecological efficiency, the fewer the lead resources consumed or the fewer the lead waste and pollutants emitted. Conversely, it means that additional services can be achieved under the same environmental load. By separately describing Eqs. (5.4) and (5.5), the ecological efficiency can be obtained. The laws of the relationship between the efficiency of resources and environmental efficiency change with the parameters and can be obtained separately.

5.3.1 Resource Efficiency

The definition of the ratio of the product yield is $P_{\tau-\Delta\tau}$ prior to year $\Delta\tau$, and the product yield P_τ in the current year is the change ratio of the product yield expressed as p:

$$p = \frac{P_{\tau-\Delta\tau}}{P_\tau} \tag{5.6}$$

p is called the lead-acid battery output change ratio, reflecting the battery output changing with each year; the value is always positive.

Substituting Eqs. (5.1), (5.3) and (5.6) into Eq. (5.4) can be collated as follows:

$$r = \frac{F_P \cdot \Delta\tau}{1 + \gamma - \alpha p} \tag{5.7a}$$

To more clearly observe the changing relationship between resource efficiency and annual growth rates, it is assumed that the yield of lead-acid batteries increases linearly with the annual growth rate ρ:

$$\frac{P_{\tau-\Delta\tau}}{P_\tau} = 1 - \rho\Delta\tau \tag{5.8}$$

Substituting Eq. (5.8) into Eq. (5.7a) can be collated as follows:

$$r = \frac{F_P \cdot \Delta\tau}{1 + \gamma - \alpha + \alpha\rho\Delta\tau} \tag{5.7b}$$

It can be observed that the resource efficiency of lead in lead-acid batteries is a function of the lead emission rate, cycle rate, service life of lead-acid batteries $\Delta\tau$, specific energy F_P, and yield growth rate. To more clearly observe the resource

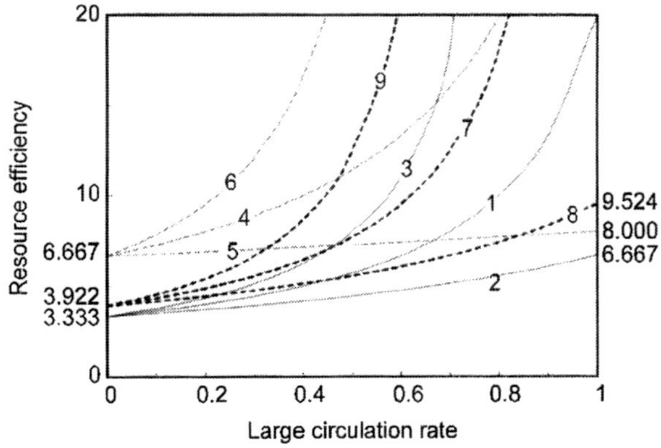

Fig. 5.3 Resource efficiency variation curves. *Source* Mao [7]

1: γ = 0.2 t/t, Δτ = 4 a, ρ = 0
2: γ = 0.2 t/t, Δτ = 4 a, ρ = 0.1
3: γ = 0.2 t/t, Δτ = 4 a, ρ = −1
4: γ = 0.2 t/t, Δτ = 8 a, ρ = 0
5: γ = 0.2 t/t, Δτ = 8 a, ρ = 0.1
6: γ = 0.2 t/t, Δτ = 8 a, ρ = −0.1
7: γ = 0.02 t/t, Δτ = 4 a, ρ = 0
8: γ = 0.02 t/t, Δτ = 4 a, ρ = 0.1
9: γ = 0.02 t/t, Δτ = 4 a, ρ = −0.1

efficiency changing with the value of parameters, by reducing some of the values of the parameters, we can obtain Fig. 5.3 based on Eq. (5.7b).

We can infer the following from Fig. 5.3:

(1) The impact of large circulation rate α on resource efficiency r—When the other parameters remain unchanged (as shown in Fig. 5.3), the efficiency of resources increases with the increase in the large circulation rate. Therefore, we must focus on improving the material circulation rate when we have a circular economy. The circulation rate can be used as a significant indicator of a circular economy.
(2) The impact of the rate of change in production ρ on the relationship between r and α—The rate of change in production will affect the speed of resource efficiency changing with the large cycle rate, and the highest value of the efficiency of resources can be achieved. Compared with constant production, the growth rate r changing with α of is slower, and the highest value of resource efficiency can be achieved when the yield continues to increase. Moreover, the faster the yield increases, the more obvious the effect. Conversely, with the declining production, the change in resource efficiency is the opposite as shown by lines 1, 2, and 3 in Fig. 5.3. Therefore, in the case of decreasing production, it is easy to obtain the significant improvement of resource efficiency.
(3) The impact of service life $\Delta\tau$ on the relationship between r and α—The service life affects the efficiency of the resource but does not affect the speed of resource efficiency changing with a large cycle rate. With all other factors held constant, the longer the life is, the higher the efficiency of resources, as shown by lines 1 and 4 in Fig. 5.3.

(4) The impact of lead emission rate γ on the relationship between r and α—The rate of lead emissions affects the speed of resource efficiency with a large cycle rate and the starting and ending values of resource efficiency. The lower the lead emission rate, the faster the growth of r with α and the greater the value of resource efficiency, as shown in lines 1 and 7 in Fig. 5.3. The opposite is also true. It is speculated that the reduction of the lead emission rate is conducive to a substantial increase in lead resource efficiency.

Finally, the efficiency of lead resources is proportional to the performance of lead-acid batteries; that is, under unchanged conditions, the higher the performance, the higher the efficiency of resources. This is simple to observe from Eq. (5.7).

It can be concluded that if we want to improve resource efficiency, we must improve the material performance of the product (which is closely related to the "dematerialization" mentioned in the following section) and the material recycling rate, extend the service life of the product (which provides theoretical support for product maintenance and reuse in a recycling economy), and reduce the rate of material emissions in the anthropogenic system and the rate of growth of material production.

5.3.2 Environmental Efficiency

Equations (5.1), (5.3), and (5.6) are substituted into Eq. (5.5), and based on using the concept of environmental efficiency to analyze resource efficiency, we obtain the following:

$$q = \frac{F_P \cdot \Delta\tau}{\gamma + (1 - \alpha)p} \tag{5.9a}$$

In the case of a linear change in lead production, Eq. (5.8) is substituted into Eq. (5.9a), resulting in the following:

$$q = \frac{F_P \cdot \Delta\tau}{\gamma + (1 - \alpha)(1 - \rho\Delta\tau)} \tag{5.9b}$$

The symbols have the same meaning as in the previous equations.

It can be seen that the environmental efficiency of lead q is also a function of the specific performance, service life, lead emission rate, lead large cycle, and yield growth rate of lead-acid batteries, which also means that there is a relationship between environmental efficiency and resource efficiency.

5.3.3 Relationship Between Resource Efficiency and Environmental Efficiency

To analyze the quantitative relationship between resource efficiency and environmental efficiency, combining Eqs. (5.7a) and (5.9a), we obtain the following:

$$\frac{1}{q} = \frac{1}{r} - \frac{1-p}{F_P \cdot \Delta\tau} \tag{5.10a}$$

In the case of a linear change in lead-acid battery production, substituting Eq. (5.8) into Eq. (5.10a) results in the following:

$$q = \left[\frac{1}{r} - \frac{\rho}{F_P}\right]^{-1} \tag{5.10b}$$

To more clearly observe environmental efficiency changing with resource efficiency, by setting some of the values of the parameters, we obtain Fig. 5.4 based on Eq. (5.10b).

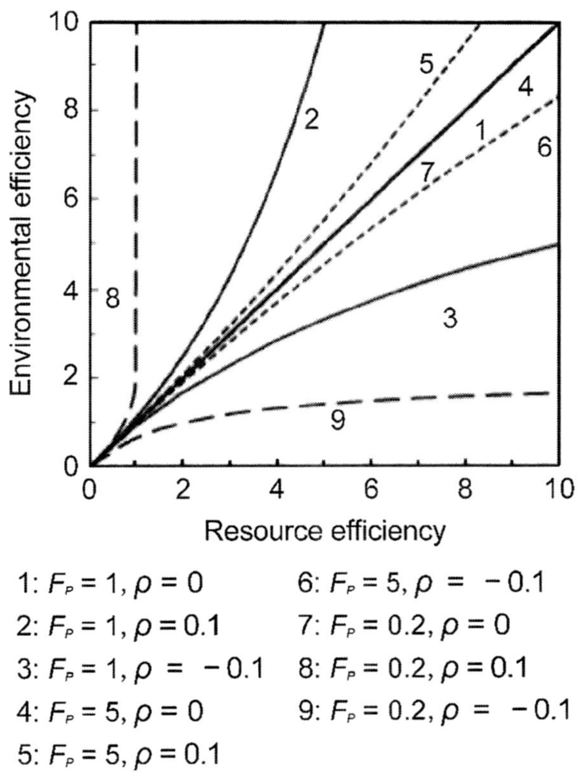

Fig. 5.4 Curves of environmental efficiency changing with resource efficiency. *Source* Mao [7]

As shown in Fig. 5.4, in all cases, environmental efficiency increases with an increase in resource efficiency, and the environmental efficiency is equal to the resource efficiency in the case of constant production, represented as a straight line in slope 1.

Further analysis shows that the changing speed of environmental efficiency with resource efficiency will be affected by the rate of change in production and ratio performance. A change in production will affect the relative values of environmental efficiency and resource efficiency. In the case of production growth, environmental efficiency will be constantly greater than resource efficiency; conversely, in the case of declining production, environmental efficiency will be constantly less than resource efficiency. Under the same yield change rate, the output growth and the decline in production are precisely inclined to form a symmetrical relationship, as shown in Fig. 5.4, lines 2 and 3. It is speculated that under the same resource efficiency, yield growth contributes to the improvement of environmental efficiency. The ratio performance will primarily affect the environmental efficiency changing with the resource efficiency. In the case of continuous growth in production, as shown in Fig. 5.4, lines 2 and 5, if the ratio of performance is higher, line 5 is below line 2, indicating that the speed of environmental efficiency increasing with resource efficiency is slower, and the higher the performance, the lower the environmental efficiency if the resource efficiency remains the same. However, in the case of a continuous decline in production, the impact of ratio performance on environmental efficiency and resource efficiency is the opposite. Regardless of growth or decline in production or the performance ratio, the higher the efficiency of resources, the higher the environmental efficiency, but the degree is different. Therefore, we can use improving resource efficiency as the focus when we are in a circular economy.

5.4 Another Expression of the Law of Man-Made Circulation of Material

As mentioned previously, the volume of service of a system can also be expressed as the number of products provided to society, that is, the product output during the statistical period. If the yield of the products in the statistical period is selected to represent its service volume for the lead-acid battery systems in the previous section, which is calculated by its lead content, the lead volume in the system will be expressed as follows:

$$S' = P_{\tau} \tag{5.11}$$

The resource load and environmental load of the system continue to be expressed as in Eqs. (5.1) and (5.2). The units of these parameters are t/a.

Similarly, Eqs. (5.1), (5.2), and (5.11) are substituted into Eqs. (5.4) and (5.5). If Eq. (5.6) is considered, we can obtain another expression of the ecological efficiency that uses the product quality of the product as the service volume:

Resource efficiency:

$$r' = \frac{1}{1 + \gamma - \alpha p} \tag{5.12a}$$

Environmental efficiency:

$$q' = \frac{1}{\gamma + (1 - \alpha)p} \tag{5.13a}$$

In this case, if the relationship between Eqs. (5.12a) and (5.13a) is established, the relationship between environmental efficiency and resource efficiency can be solved:

$$\frac{1}{q'} - \frac{1}{r'} = p - 1 \tag{5.14a}$$

It can be observed that in the case of constant production, resource efficiency is equal to environmental efficiency. In the case of increasing production, resource efficiency is constantly less than environmental efficiency. In the case of declining production, resource efficiency is constantly greater than environmental efficiency. This is primarily due to changes in product yield, affecting the product system upstream and downstream. Because of production growth and system expansion, availability of a large quantity of resources, and few emissions, resource efficiency is less than environmental efficiency; however, a decline in output, a reduction in system size, limited availability of resources, and a large quantity of emissions cause the resource efficiency to be greater than the environmental efficiency.

If the linear change of the yield can be further considered and Eq. (5.8) substituted into Eqs. (5.12a) and (5.13a), Eqs. (5.12b) and (5.13b) can be obtained by collation:

Resource efficiency:

$$r' = \frac{1}{1 + \gamma - \alpha + \alpha\rho\Delta\tau} \tag{5.12b}$$

Environmental efficiency:

$$q' = \frac{1}{\gamma + (1 - \alpha)(1 - \rho\Delta\tau)} \tag{5.13b}$$

Compared with Eqs. (5.7) and (5.12), (5.9) and (5.13), it can be seen that the variation of resource efficiency and environmental efficiency does not show the ratio performance of products when the number of services is characterized by the quantity of products. That means the benefits get from material reduction and weight light cannot be shown; for example, after the miniaturization of the computer, the service function of a certain computer is being strengthened, and the amount of the consumed substance is decreasing and the performance is greatly improved, which is submerged in the research method.

Aa an extension, we recommended students read the papers [8, 9]. Consider what type of resource efficiency and environmental efficiency expression can be derived under the logistics framework in the article "Iron Flow Analysis of Iron and Steel Product Life Cycle." Compared with the equations provided in this section, students should analyze the differences and the reasons for the formation of these differences. Consider which is more scientific, allowing students to experience that a scientific research course is the deepening process of research methods, results, and conclusions.

5.5 The Ideal Model of Man-Made Material Circulation

In the man-made flow of material, if its product system does not consume mineral resources and does not release any of the material containing waste and pollutants into the environment, the ideal material cycle will be formed in its product system. In this section, lead continues to be used as an example. If the values of Eqs. (5.1) and (5.2) are zero and Eq. (5.6) is considered, the following equation will be collated:

$$\begin{cases} p = 1 \\ \alpha = 1 \\ \gamma = 0 \end{cases} \tag{5.14a}$$

If the annual rate of change in output continues to be expressed as ρ, then the following:

$$\begin{cases} \rho = 0 \\ \alpha = 1 \\ \gamma = 0 \end{cases} \tag{5.14b}$$

Equation (5.14) shows that in the case of ideal circulation and the yield (material content) of the products at the end of each year remains unchanged, the increase in social service will be manifested as an increase in resource utilization rate caused by material reduction. After being scrapped, all products are to be recycled and all the material is recycled. No waste or contaminants are emitted during operation of the product system. Equation (5.14) is the ideal circulatory equation of material that is a necessary and sufficient condition to realize the ideal mode of man-made material flow.

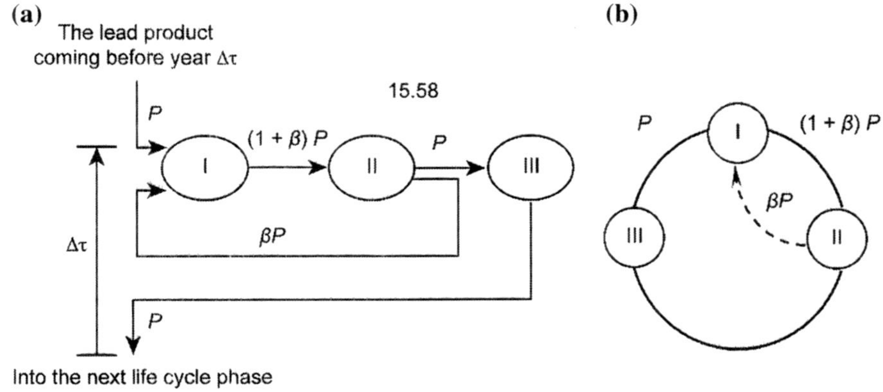

Fig. 5.5 Ideal circulation flow of matter: **a** the service life $\Delta\tau$ is visible; **b** the service life $\Delta\tau$ is invisible. *Source* Mao [7]

Based on the above findings, we must improve the material circulation rate and reduce the waste discharge to realize the ideal man-made flow of material, so that all the waste must be recycled, including the waste and products produced in the production process that have completed their service lives and have become depreciation waste. After these changes, the man-made flow of material will be close to the ideal circulation based on stable material content of year-end products.

Under the ideal circulation flow condition, Fig. 5.1 is simplified to Fig. 5.5.

By comparing Figs. 5.5 and 5.1, it is simple to observe that the connection between the input of material mineral resources and the discharge into the waste environment in Fig. 5.1 disappears, indicating that the entire system no longer obtains any mineral resources from the environment, and concurrently, no waste or contaminants are discharged into the environment. The system completely relies on the material stored within itself to continue production.

In addition, the scientificity of analysis of the model, the research method, and the conclusion are further verified by the mathematical analysis of the ideal circulation flow of material.

5.6 Method to Apply a Case Analysis

Because the material flow of human beings has specific laws, management practice should follow these laws to continuously optimize the material flow of mankind. These laws are primarily used from the following two aspects. First, an effective indicator system can be established according to the basic law, including the quantitative assessment of indicators applied to integrated management in human development, resource conservation, and environmental improvement. Second, with the analysis of the status of material flow, we can predict the future and employ effective management strategies based on the law. According to the assessment of indicators and their application and case analysis and improvement, we analyze cases in the following sections.

5.6.1 Application Example 1: Establishing Evaluation Indicators and Incorporating Them into Management Applications

Whether it is a resource and environment system or a human socio-economic system, its management requires quantitative management indicators. Based on analysis of the human flow of material, it is not difficult to determine that the human flow process involves many different indicators. These indicators can be divided into absolute indicators and relativity indicators. The absolute indicators reflect the size of the material flow, and the relativity indicators reflect the efficiency of resource use or the ratio of material recovery to pollutant emissions. Additionally, based on the relationship between the product system and the resources and environment, indicators can be divided into internal indicators and external indicators of a product system. External indicators refer to the influential indicators of the product system and the resource and environment, while the internal indicators refer to the material link indicators of all subsystems or the life cycle stages in the internal product system. The indicators for evaluating the man-made flow of material and their implications are summarized in Table 5.2.

5.6.2 Application Example 2: Practical Application Validation and Applying Management Measures

The lead-acid battery system research in the previous chapter continues to be used as the example.

To find a method to improve the lead flow of lead-acid battery systems in China, the resource efficiency, environmental efficiency, and values of various parameters were calculated based on the results of the lead flow analysis (Fig. 5.2). Comparing the values from China with other countries that have researched similar products, we can obtain the lead flow status of lead-acid battery systems in different countries, as shown in Table 5.3.

As shown in Table 5.3, the resource efficiency and environmental efficiency of lead in lead-acid batteries are as high as 15.804 (MWh) a/t, which is 132.92 times and 120.66 times that of China. This is because in terms of circulation rate, Sweden has reached 0.99 t/t for almost all cycles, while China is only 0.312 t/t. In terms of lead emissions, Sweden's γ is only 0.002 655 t/t, while China's is as high as 0.324 t/t. In terms of yield changes, Sweden had stabilized the life cycle of a lead-acid battery to a minimum of five years, while China's continues to grow. In terms of battery service life, Sweden can reach five years, while China has an average life of three years.

Table 5.2 List of indicators for anthropogenic material flow assessment

Classification	Target categories	Indicator examples	Meaning
Expressions based on the numeric value	Absolute indicators	Resource consumption	The amount of natural resources consumed by a product system during a statistical period
		Pollution emissions	The amount of waste and pollutant discharged from the product system into the environment during a statistical period
		Amount of social services	The amount of service provided by the product system to society during a statistical period
	Relativity indicators	Resource efficiency	The amount of services provided to the community by the natural resources consumed by the product system or the ratio of product system social service volume and its natural resource consumption during a statistical period
		Environmental efficiency	The ratio of the service volume provided by the product system to society and the quantity of waste and pollutants discharged into the environment during the same period or the social volume when the product system is discharging environmental waste and pollutants during a statistical period
		Cycle rate	Includes large and small cycle rates. Large circulation rate refers to the product after scrapping compared to the proportion of substances in the material regeneration phase, and the small cycle rate refers to the ratio of the number of materials that were generated as waste in the product processing and manufacturing stage compared to the material recovery stage and the final product output during the same period
		Waste rate	The ratio of the amount of environmental waste and pollutant loading to the final product output during a statistical period
According to the relationship between the product system and the outside	External indicators	Resource exploitation	The number of natural resources in a statistical system for a product system
		Waste emissions	The amount of waste and pollutant loading discharge to the environment during a statistical period
		Trade rate	Corresponds to the fixed-point logistics analysis. The proportion of product output from a product system during a stage
	Internal indicators	Cycle rate	Includes large and small cycle rates. Large circulation rate refers to the product after the scrapping compared to the proportion of substances in the material regeneration phase, and the small cycle rate refers to the ratio of the number of materials that were generated

(continued)

Table 5.2 (continued)

Classification	Target categories	Indicator examples	Meaning
			as waste in the product processing and manufacturing stage compared to the material recovery stage and the final product output during the same period
		Accumulation rate[a]	In-use stock corresponds to fixed-point logistics analysis. It refers to the ratio of the quantity of in-use stock of net inflow during a statistical period of a product system and the quantity of substances put into use during the same period
		Waste rate	The same as above, or the social volume when the product system discharges environmental waste and pollutants during a statistical period
		Waste recovery	The number of materials that can be recovered from all types of waste generated by a statistical product system
		Product life	The time difference between the use of the product and its scrap
According to the relationship between the product system and the outside	Internal and external contact indicators	Resource efficiency	The service volume provided by the natural resources consumed in the product system to society or the ratio of service volume of the product system and its natural resource consumption during a statistical period
		Environmental efficiency	The ratio of the service volume provided by the product system to society and the quantity of waste and pollutants discharged into the environment during the same period or the social volume when a product system discharges environmental waste and pollutants during a statistical period
		Resource assurance period	The number of years that can be used to ensure that the current resource reserves are estimated based on current mineral exploitation
		Man-made flow exceeding multiple of pollutants	The ratio of the human flow rate (anthropogenic emissions per unit of time) of a certain environmental pollutant and its natural flow rate (the number of natural cycles per unit of time)

[a]See Chap. 6 for the analysis of fixed-point logistics

Table 5.3 Comparison of lead flow in lead-acid battery systems in China and Sweden

	Resource efficiency [(kWh) a/t]	Environmental efficiency [(kWh) a/t]	Large cycle rate (t/t)	Middle cycle rate (t/t)	Lead emission rate (t/t)	Average annual growth rate (t/t)	Service life (a)
China	118.90	130.98	0.312	0.039	0.324	0.032	3
Sweden	15804	15804	0.99	0.1236	0.002655	0	5
Sweden/China	132.92	120.66	3.173	3.169	0.0082	0	1.667

Note In the specific performance of lead-acid batteries, China estimates FP = 41.303 kWh/t and Sweden estimates 40 kWh/t

According to the variation in resource efficiency and environmental efficiency mentioned previously, to improve China's lead-acid batteries in lead's ecological efficiency, we should focus on improving the lead cycle rate and reducing lead emissions. For the former, we need to employ the following measures:

(I) Learning from foreign advanced management experience, we establish waste lead recycling legislation, improve the recycling mechanism, and control and lead the recycling of waste lead batteries into the normal track.

(II) We should extend the corporate responsibility of each lead-acid battery factory, gradually realize the business transition from product-oriented to service-oriented sales, and promote the recycling of lead batteries.

(III) We should strengthen environmental protection publicity, establish the concept of "waste is also a resource," and charge levies to battery consumers and emission taxes to waste lead discharge departments to ensure the smooth progress of waste lead recovery.

To reduce lead emissions, the following must occur:

(I) Implement a special licensing management system for lead production enterprises. The scale of production, technology, and environmental protection measures of production are strictly regulated. Small mining enterprises are firmly prohibited, and small smelters are banned and closed.

(II) We can develop new mineral processing technology to improve the yield of mineral processing. We can promote an advanced non-polluting smelting lead process; eliminate small, antiquated lead smelting processes; and develop preferential economic policies to encourage new technology for smelting lead and the promotion and application of this new technology.

(III) Via publicity and education, we should promote the harmfulness of lead and methods of preventing lead pollution, form good habits for environmental protection, and strive to reduce the amount of waste lead in the environment, thus avoiding harm to the human body.

History shows that China has moved from pollution to forced management. We hope that we can learn from this experience and are able to avoid pollution by scientific management before additional pollution is formed.

5.6.3 Application Example 3: The Sources of Logistics Analysis Data

Effective and sufficient data are the prerequisites and working basis for obtaining effective research results, which have the same status as research methods in scientific research. For different research topics, the questions to be answered and the required data are different, so the data sources and calculation methods are not the same. China's lead-acid battery system logistics analysis is used as an example; its primary sources of data are shown in Table 5.4.

Table 5.4 Primary data sources for lead flow analysis in China

Data type or name	Data sources	Compilation organization
Mining yield	Investigation Report on Development and Utilization of Lead and Zinc Mineral Resources in China	Beijing Mining and Metallurgy Research Institute
Beneficiation and smelting yield	1990–2001 China Nonferrous Metals Industry Yearbook	China Nonferrous Metals Industry Yearbook Editorial Board
Regeneration amount of waste lead and production data	Published literature	Portions provided by the China Renewable Metal Association
Lead production data for acid batteries	Environmental Assessment Report of Shenyang Battery Co., Ltd	Shenyang Environmental Science Research Institute
Lead-acid battery product structure and performance	China Accumulator Annual Statistical Report	Shenyang Institute of Accumulator
Lead-acid battery production data	China Machinery Industry Yearbook China Electrical Appliance Industry Yearbook	China Machinery Industry Yearbook Editorial Board
Import and export data for lead-acid batteries	China Foreign Economic and Trade Yearbook	China Foreign Economic and Trade Yearbook Editorial Board
Import and export data for waste lead	China Foreign Economic Yearbook	China Foreign Economic Yearbook Editorial Board

Classroom Discussions and Assignments

Classroom Thinking and Discussion

The assumptions made in the study may change, or the product being studied may not conform to the assumptions of the study. For example, production and production time use the same scale and aluminum for the production of cans, but cans from production to being scrapped may occur over a few months' time. Additionally, how is a research framework built for products that are constantly depleted during their use, such as the lead in a pencil being depleted in the process of the pencil's use?

For complex substances (or systems), how are logistics analyzed?

Is there another method if we do not use the tracking logistics method?

Classroom Issues

What type of material is the most worthy of attention? Students choose a substance of interest, create an outline of its material flow, and share it in class.

Group Assignments

Based on the material selected in the discussion of classroom issues, students present classroom reports with the assistance of related logistics analysis papers located in the library literature system. The students are required to reflect the group selection process in the report and describe how they established the thesis topic, describe the research basis, and present the contents of their report. A logical relationship between the sections and paragraphs of the paper should be reflected. The report time is 10–12 min. The scoring criteria are shown in Table 5.5.

Table 5.5 Classroom group report rating table

Report title	Whether the selection process is clear (maximum 10 points)	Whether the subject matter is clear (maximum 10 points)	Whether the research basis is sufficient (maximum 10 points)	Whether the logical relationship is clear (maximum 10 points)	Whether the expression is reasonable (maximum 10 points)

Recommended and Referenced Literatures

1. Mao, J.S., Z.W. Lu, and Z.F. Yang. 2006. Eco-efficiency of lead in China's lead-acid battery system. *Journal of Industrial Ecology* 10 (1/2): 185–197.
2. Zhongwu, Lu. 2002. Iron-flow analysis for the life cycle of steel products-a study on the source index for iron emission. *Acta Metallurgica Sinica* 38 (1): 58–68.
3. Zhongwu, Lu. 2003. Study on some aspects of circular economy. *Research of Environmental Science* 16 (5): 1–10.
4. http://www.factor10-institute.org/publications.html.
5. Zhongwu, Lu. 2005. Some opinions on further planning of circular economy. *Environmental Protection* 1: 14–17.
6. Zhongwu, Lu. 2006. The following-observing method for substance flow analysis. *Engineering Science* 8 (1): 18–25.
7. Jiansu, Mao. 2003. *Industrial metabolism of lead and its influence on national economy*. Shenyang: Northeastern University.
8. Zhongwu, Lu. 2002. Iron flow analysis of iron and steel product life cycle. *Acta Metallurgica Sinica* 38 (1): 58–68.
9. Zhongwu, Lu. 2003. Study on several problems of circular economy. *Environmental Science Research* 16 (5): 1–10.

6 Circulation Flow of Material: Fixed-Point Logistics Analysis

6.1 Presentation of Core Issues

There are two representative methods of material flow analysis: the tracking method that focuses on the material particle and the fixed-point method that focuses on a particular region. The previous chapter explored the tracking method of logistics analysis. This chapter will focus on the fixed-point method combined with the Yale University logistics analysis framework to study the evolution of its scientific research process.

This chapter is primarily used to answer questions, including the following: How is the material of a study selected? How is a man-made circulation flow of materials analyzed? What are the social and environmental effects of these flows? What are the uses of these studies? Finally, we present a number of thoughtful questions to allow students to expand their thinking and to choose from for use in relevant research.

6.2 Method to Select a Substance for Study

At the Stocks And Flows (STAF) group workshop held at the Yale University Industrial Ecology Research Center in 2006, Graedel shared the principles of the team's substance screening.

(I) Substances that cannot be circulated naturally—The group quantitatively reviewed the natural circulation level and the degree of human disturbance of each substance in the periodic table, from which a class of substances whose anthropogenic disturbances are far greater than their natural circulation levels was selected as the focus of the material type.

(II) Material with a higher use rate—The use rate refers to the annual usage of a material. The higher the rate of use, the more important the material for

J. Mao et al., *Circular Economy and Sustainable Development Enterprises*,
https://doi.org/10.1007/978-981-10-8524-6_6

human beings. Additionally, material shortages are more likely; therefore, we should focus on these materials.

(III) Material with a faster use rate of change—The speed of change of a use rate reflects the amount of material usage changes per unit period, which reflects the relative relationship between human activities and external resources. The more rapidly the index changes, the easier it is to change the relationship between human beings and the environment. For example, the faster the rate of use increases, the more depleted the material resources.

(IV) Highly toxic and highly harmful material—This property indicates the related potential toxic risk levels of the material in the environment. Particular attention should be paid to material that is toxic and harmful to human health.

(V) Material related to science or technology—This property is used to indicates the potential usage in scientific research or technology; for example, a class of material may currently be depleted, which indicates the need to increase the level of recycling in scientific research or find suitable alternative substances, such as copper, platinum, or zinc; another example is the accelerated use of substances, such as in many cases began to use aluminum instead of copper on many occasions which led to a rapid increase in the amount of aluminum each year; and there are some closely related to other substance recycling processes such as brass, solder, stainless steel, and other production that contains iron, tin, nickel, and many other different substances.

(VI) Materials that are funded by scientific research—This condition means that other sectors are also interested and are willing to provide financial support for related research such as the National Natural Science Foundation, professional and technical associations, or a specific organization's advisory project.

6.3 Using the Fixed-Point Method to Analyze Material Flow

6.3.1 The Logistics Analysis Framework of the Fixed-Point Method

In the logistics analysis framework of the fixed-point method, researchers will focus on specific research areas, such as the world, a country, or another specific area. Currently, metal elements are the research object in the majority of research.

The logistics analysis framework of the fixed-point method is shown in Fig. 6.1. The material flow process includes four stages: metal materials production, metal products processing and manufacturing, the use of metal products, and the recycling and management of waste when products have completed their service lives. The crustal lithosphere serves as the initial source of metal elements, while the human external environment, such as landfills, sedimentation tanks and roads and their structures, which have become the locations where metal elements have been

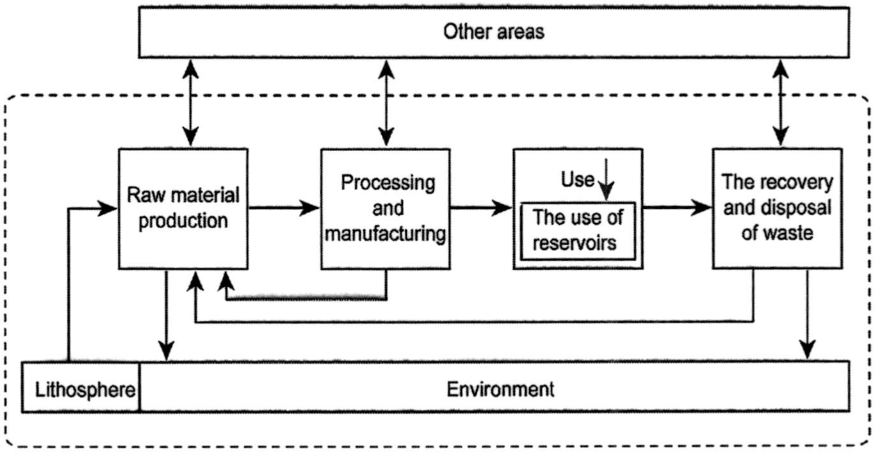

Fig. 6.1 STAF basic framework of material flow reprinted from (Mao et al. [7])

excreted as waste and pollutants, is the final location for the metal element as environmental waste. During the use of metal products, metal elements will be accumulated in the human social system, forming an in-use stock of metal elements. The law of conservation of matter is observed between material element inflows, outflows, and inventory changes for each life cycle period. The continuity of material flow constitutes the flow of material elements among stages.

In application, Fig. 6.1 can be further refined due to changes in metal species, data availability, research objectives, and other factors. For example, when the model framework is used to study the cyclic flow of lead elements, we should consider that contaminants containing lead may threaten the environmental ecosystem, especially the health of the human body; therefore, special attention must be focused on the emission of pollutants containing lead. Consequently, the manufacturing stage and the emission of pollutants containing lead in the use stage are to be supplemented based on the original framework (Fig. 6.1). Additionally, because lead waste from different sources has different environmental risks, subscripts are used to distinguish the source of the lead contaminants, as shown in Fig. 6.2.

The method to apply the framework in Fig. 6.1 to analyze the man-made flow of material elements was created by Graedel of Yale University, and it has been widely used in the flow research of multiscale cycles, such as copper, steel, zinc, and other metal materials. It is called the fixed-point method of material flow analysis by academician Lu Zhongwu.

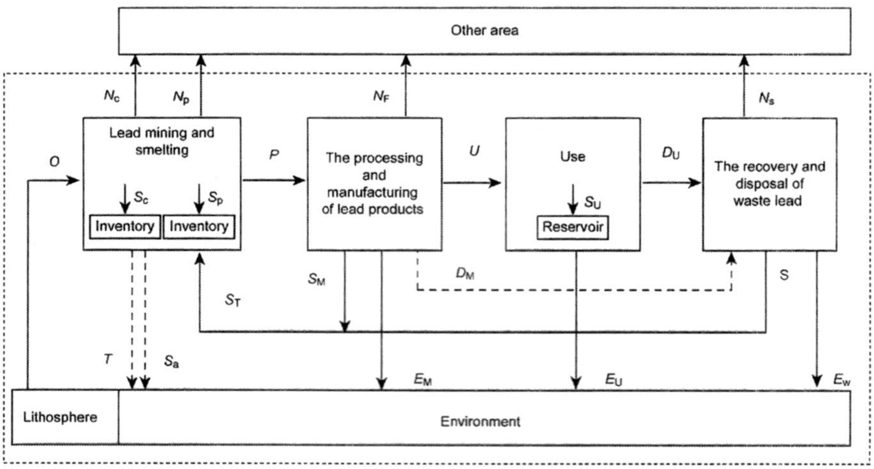

System Boundaries © STAF Project, Yale University, 2006.

Symbols and definitions

O: Lead ore resources
P: Refined lead
U: Lead products being placed into use
D_U: The depreciation of waste lead entering into the waste lead recovery and disposal phase
D_M: The processing of waste lead in the waste lead recovery and disposal stage
N: Net exports (export volume minus import volume)
N_P: Refined lead net imports
N_C: Lead ore net imports
N_S: Waste lead net imports
N_F: Net imports of lead products and semi-finished products
T: Lead tailings
S_a: Leaded slag
S_P: Refined lead inventory
S_C: Refined ore stock
S_U: The influx of lead use accumulation
S: Waste lead resources obtained by waste lead recovery and disposal
S_T: The amount of waste in smelting
S_M: Processing waste lead
E: Waste and pollutant emissions containing lead
E_M: Emissions of lead pollutant in manufacturing and processing stage
E_U: Lead pollutant emissions in use stage
E_W: Lead pollutant emissions in waste lead recovery and disposal stage

Fig. 6.2 Man-made flow framework of lead using the "fixed-point method" lead element reprinted from (Mao et al. [7])

6.3.2 Logistics Analysis Example of the Fixed-Point Method: Lead Flow Analysis

As mentioned previous, fixed-point logistics analysis focuses on a specific spatial area. Lead continues to be used as the example.

Lead has a variety of uses, including being used to produce a variety of lead products. Different lead products' processing and manufacturing and scrap recycling processes are complex and various based on different products, so the lead flow in Figs. 6.1 and 6.2 will be further refined and embodied in each stage of the lead product life cycle, as shown in Fig. 6.3a–d.

Based on statistical data and a variety of technical data from the production process, the lead flow in different spatial scales can be categorized as shown in Fig. 6.3 and summarized into a graph to obtain a lead flow diagram for a particular year. Figure 6.4 is a detailed diagram of the lead flow in North America in 2000,

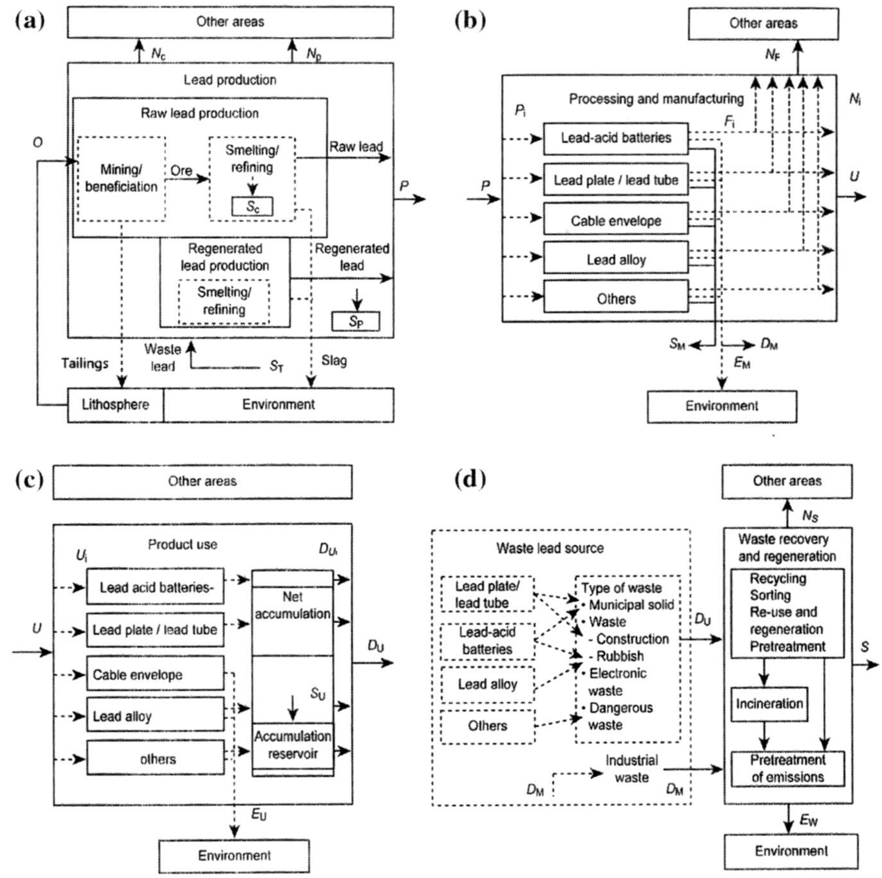

a) Lead metal production; b) The processing of lead products; c) The use of lead products; d) Waste lead recovery and regeneration

Fig. 6.3 Detailed analysis of lead flow during each life cycle stage of the man-made flow of lead reprinted from (Mao et al. [7])

showing a variety of uses of material elements and the import and export trade factors of a variety of products and semi-finished products.

In application, the lead flow of a particular area can also be plotted, as shown in Fig. 6.2. Figure 6.5 shows the global lead flow diagram for 2000. Figure 6.6 shows the lead flow diagram for China in 2000. There is no trade flow in the lead flow diagram at the global level, and the lead flow in a particular region will have a trade flow with other regions. Figures 6.5 and 6.6 show the interference intensity of the man-made lead flow on the natural system in 2000 at the global and Chinese levels, which can be used as an important basis for a lead ore resources protection policy for a specific area, the source control of lead pollution, and the implementation of a waste lead recycling economy.

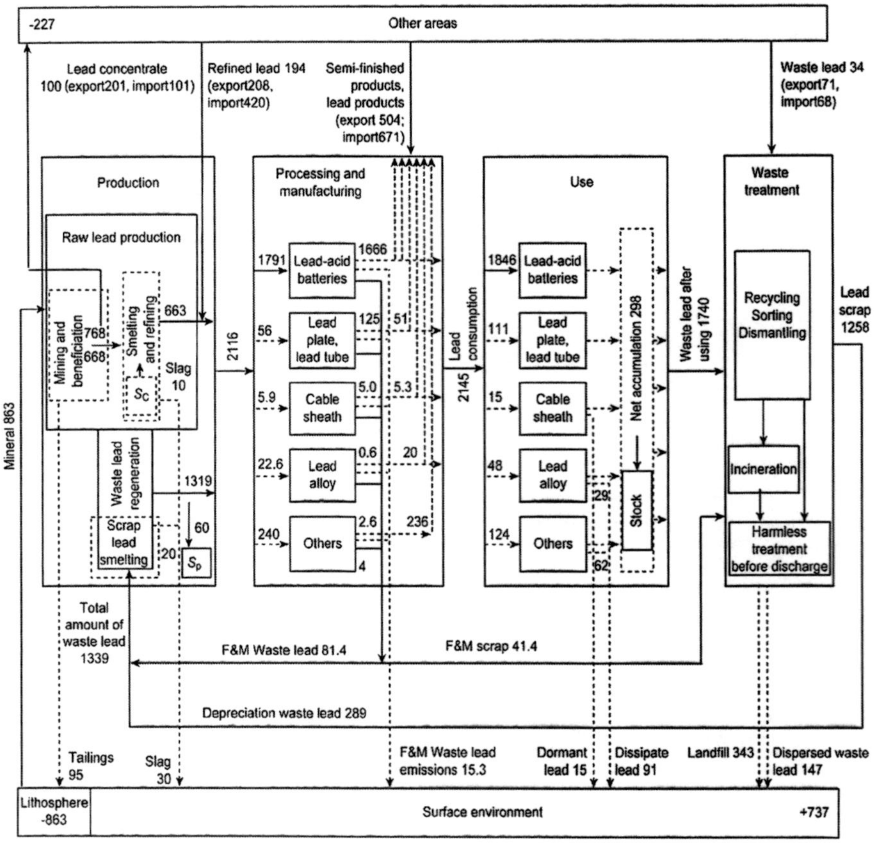

Fig. 6.4 Detailed diagram of the man-made flow of lead elements in North America in 2000 (unit: Gg/a) reprinted from (Mao et al. [8])

In addition, if we compare Figs. 6.2 and 6.6, we can observe the main differences in the lead flow diagrams between the tracking and fixed-point methods.

6.4 What Is the Result of Man-Made Flow? Example 1: Social Effects

6.4.1 What Is the Use Accumulation of Material?

In-use stock refers to the sum of all the substances of all types of services that are currently being offered to human beings, referred to as the "in use" of a substance. When the amount of material placed into use during a period is greater than the amount of material that is not used, the difference will accumulate in the "use"

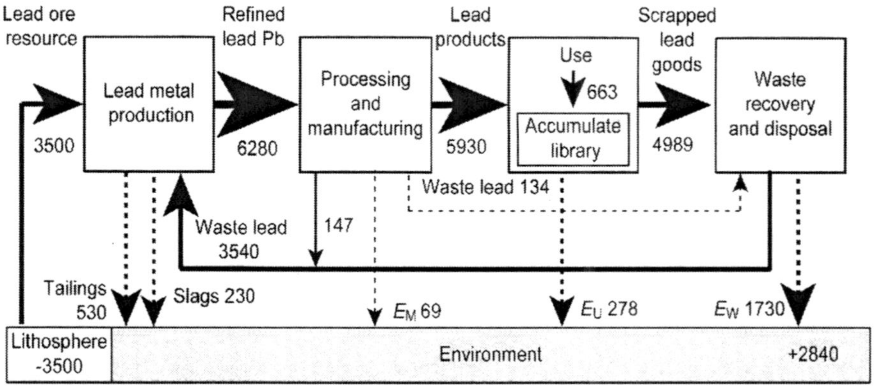

Fig. 6.5 Chart for the man-made flow of the global lead elements in 2000 (unit: Gg/a) reprinted from (Mao et al. [8])

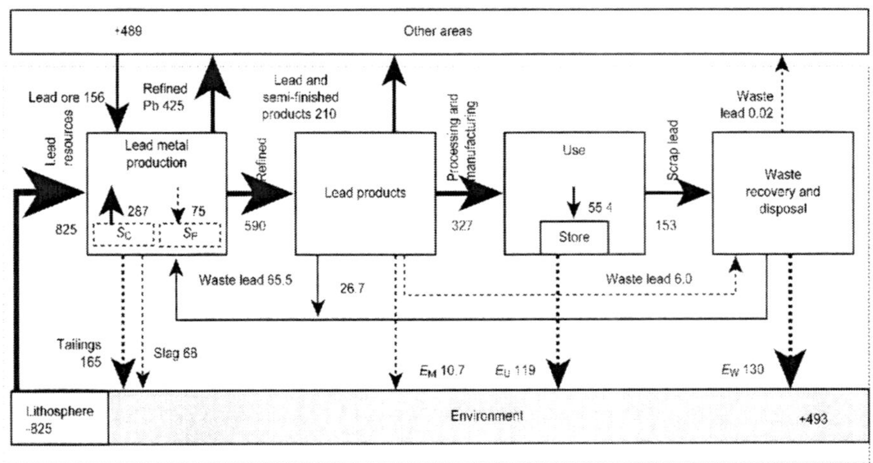

Fig. 6.6 Diagram of the man-made lead flow in China in 2000 (unit: Gg/a) reprinted from (Mao et al. [8])

phase of the material, resulting in an increase in the amount of material. The continued increase in the amount of material in use will result in the accumulation of material in the human social service system, the formation of an anthropogenic and secondary location for the storage of substances in service to meet human needs or being used by human beings, referred to as the in-use stock of material. Therefore, "in-use stock" can refer to the amount of substances being used but also refers to the storage locations for substances that are collectively in a use state.

The amount of material in-use stock in a country or region can reflect the standard of living under the specific technology of the country or region; therefore, it can be used as an indicator of the quantity of material needed for less developed countries or regions to achieve the same standard of living as the developed countries. Additionally, the current amount of material in in-use stock will leave use in the future when the products are scrapped, so the amount of material in-use stock that can reflect the amount of waste is important data for waste management and recycling.

6.4.2 Method to Estimate the Use Accumulation of Material

The net inflow of the in use of lead in each year can be obtained by analyzing the annual lead flow in a country or region. Then, we can sum the amounts of the net increase of each year during a given historical period to obtain the total amount of in-use stock during the historical period; however, this requires a great deal of effort. Therefore, we propose a simpler method.

There are two methods to estimate the amount of materials in use. One is to calculate the difference between the amount of materials put into use and the amount leaving use, and sum these amounts during a specific historical period by means of material flow analysis. This is called the top-down approach. The other method calculates the amount of each service unit containing a specific substance and analyzes the content of the substance in each service unit and then calculates the amount of the substance in-use stock, which is called the bottom-up approach. The following sections show the calculation processes of these two methods using lead as an example.

(I) Estimation Method 1—Top-down Approach

As discussed above, the top-down approach is used to estimate the amount of materials in use by employing material flow analysis. The lead flow diagrams in the material use phase described in Figs. 6.2 and 6.3 may be separately analyzed.

First, we analyze the calculation of the number of single lead product systems in use.

As can be observed from Figs. 6.2 and 6.3c, the net increase in the material "in use" for year τ is the amount of material U_τ placed into use for that year minus the amount of material not used in that year. The amount of substances leaving during that year includes the following two flows: the depreciation waste D_U due to end of product life and waste lead and lead contaminants E_U gradually dissipating into the environment over the course of the product use due to corrosion, wear, and other reasons. Thus, the net inflow of "the use of the repository" for the year can be expressed as:

$$S_{U\tau} = U_\tau - D_{U\tau} - E_{U\tau} \tag{6.1}$$

where $S_{U\tau}$ is the net amount of lead in a certain lead product system during the year, also called the net inflow of lead in use in year τ.

Assuming that the service life of the lead product is $\Delta\tau$ and the environmental release during the use of the product occurs when the product is scrapped, it is not difficult to envisage that the amount of materials to be used in the year should be equal to the amount of materials placed into use prior to year $\Delta\tau$:

$$D_{U\tau} + E_{U\tau} = U_{\tau-\Delta\tau} \tag{6.2}$$

Substituting Eq. (6.2) into Eq. (6.1), we obtain the following:

$$S_{U\tau} = U_\tau - U_{\tau-\Delta\tau} \tag{6.3}$$

It can be seen that for a single product system, the annual net increase in the amount of materials in use in a given year is the difference between the amount of materials placed into use in that year and the amount of materials placed into use in a previous life cycle year.

If a specific period T is selected, the net increase in the amount of the materials in use will be the cumulative result of the net increase in the material for each period of the year, which can be expressed as follows:

$$\sigma \equiv \int_0^T S_U \mathrm{d}t \equiv \sum_{\tau=0}^{T} S_{U\tau} \tag{6.4}$$

In fact, a substance usually has a variety of different uses, so multiple product systems are involved. Lead is no exception. For composite systems with a variety of different product systems, the subscript i can be used to distinguish the different product types. For example, lead products are divided into five types as shown in Fig. 6.7, where batteries are divided into starting, lighting, and ignition (SLI) and fixed batteries. While the service life of different types of products, the stability of the substances in the use process, and the performance of the recyclability of the waste are very different, they should be treated differently when the amount of lead in use is estimated. In this case, the amount of lead placed into use in a given year will be the sum of the lead content of the various types of lead products that will be used in that year:

$$U = \sum_{i=1}^{N} U_i \tag{6.5}$$

Substituting Eq. (6.5) into Eq. (6.3), we can obtain a net increase in the amount of in-use stock of a variety of lead products for a given year:

$$S_{U\tau} \equiv \sum_{i=1}^{N} S_{U\tau i} = \sum_{i=1}^{N} (U_\tau - U_{\tau-\Delta\tau_i})_i \tag{6.6}$$

Substituting Eq. (6.5) into Eq. (6.4), we can obtain the total amount of materials in use in a given period of time for a variety of lead products:

$$\sigma \equiv \sum_{\tau=0}^{T}\sum_{i=1}^{N}(U_{\tau} - U_{\tau-\Delta\tau_i})_i \tag{6.7}$$

In Eqs. (6.5) through (6.7), N is the number of product types involved.

The above analysis shows that the top-down method for estimating the amount of material in-use stock will involve the material consumption structure (or product type), the service life of various products, the increase and decrease in products' annual outputs, the historical period of inspection, and other factors. It is not difficult to imagine that the in-use stock is the result of the long-term growth of material inputs. If the amount of materials placed into use is less than the amount of materials remaining during a historical period, the amount of materials in use is reduced, as is the in-use stock of materials.

(II) Estimation Method 2—Bottom-up approach

As discussed above, the bottom-up approach for estimating material in use is a method to calculate the total amount of material in use by using the amount of material in the study area. Therefore, the application of the bottom-up approach is usually divided into different types of products based on the material they use and then counting the number of various types of products in the study area and analyzing the content of each type of product; thus, the in-use stock of target materials in the study area can be calculated as follows:

$$M = \sum_{i=1}^{N_T} C_i \cdot N_i \tag{6.8}$$

where N_i is the number of i-type products (unit is piece); C_i is the target material content in i-type products (unit is kg/piece); N_T is the number of product types of materials; and M is the amount of materials in use (unit is kg).

If the study area is large, it can be divided into several regional units and the amount of the materials in use is calculated for each regional unit; then, the total amount of materials in use in the entire study area is aggregated.

6.4.3 Case Analysis—The Accumulation of Lead

In the top-down approach, it is necessary to specify the number of substances placed into use each year. While the material placed into use is implied in a variety of different products, the number of different products in each year is usually difficult to obtain, affecting the availability of data to support the method. Therefore, it is desirable to use data that are more readily available for estimation.

In the case of lead, because it is much easier to obtain lead consumption data for each year than to obtain the consumption data of lead products for each year, the work efficiency of the top-down approach for estimating the amount of material used will be greatly improved if the mathematical relationship between the consumption of lead products and lead can be established. Based on this, the two parameters of material consumption structure coefficient and material processing efficiency can be introduced; thus, the following relationship is between the amount of material placed into use and lead consumption P:

$$U_i = P \cdot f_i \cdot f_{mi} \tag{6.9}$$

where f_i represents the material consumption structure system, indicating that the fraction of the amount of lead used to make the ith lead product accounts for the total amount of lead consumption, which can be estimated by means of the consumption structure of the material; f_{mi} represents the processing efficiency rate of the material used in the production of the ith lead product, which can be obtained through engineering technical manuals or enterprise field research; and the remaining symbols are the same as in previous equations.

If the material consumption structure coefficient, processing utilization, product life, and material consumption for a given period of time are given, the amount of material stock used during a historical period can be estimated. Table 6.1 shows the lead parameter values during the process of using stocks estimated between 1995 and 2000. Figure 6.7 is a historical evolution of available lead between 1930 and 2000.

As seen from Fig. 6.7, the in-use stocks of lead in the world have increased annually for more than half a century; concurrently, its product structure is evolving, from the stereotype and lead tubes being the primary products from the

Table 6.1 Example of using the top-down approach to estimate the value of the parameters used in the storage of lead elements (2000)

	Consumption structure coefficient	Processing utilization	Service life/a
Lead-acid batteries	0.73	0.91	4 trailed models; 12 stable forms
Stereotype/lead tube	0.06	0.89	30 lead tubes; 50 stereotypes
Cable sheath[a]	0.02		
Lead alloy	0.03	0.88	10
Others	0.16	0.96	12

Source Mao and Graedel [9]

[a]Hibernating use

According to hibernating use, a cable sheath is considered to remain permanently after being placed into use

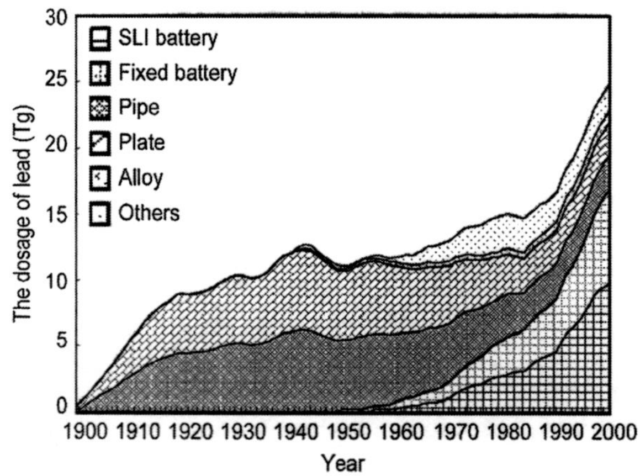

Fig. 6.7 Historical changes in the in-use stocks of lead in the world during the twentieth century reprinted from (Mao and Graedel 2009), with permission from John Wiley and Sons

middle of the twentieth century to lead-acid batteries in the second half of the twentieth century, reflecting the historical evolution of lead consumption patterns driven by human demand.

6.5 What Is the Result of the Man-Made Flow? Example 2: Environmental Effects

6.5.1 The Accumulation of Environmental Emissions in the Environment

Due to industrial metabolism, the material that was placed into production cannot be fully converted into the target product during the anthropogenic flow of material process. Those substances that cannot be converted into the target product cannot be fully recovered as a secondary resource to return to production; thus, they will form releasers that are discharged to the environment, which is referred to as environmental release, as shown in Fig. 6.2. This portion of the material from different product life cycle stages has experienced different human transformation processes; thus, it is in a different state than when it was originally exploited from the resources. It has lost its original "friendly" relationship with the natural environment and has become environmental interference factors, environmental waste, and pollutants.

According to the natural circulation of material and the anthropogenic flow discussed in Chap. 4, when the human environmental release rate exceeds the natural circulation rate of a material, the excess will accumulate in the environment. Over time, other new anthropogenic and secondary locations are formed to store those environmental releasers on the surface, referred to as the environmental release reservoirs of material. The material in the environmental release reservoirs is both the end state of human activity and the starting point of further migration into the environment. Therefore, the quantity and state of occurrence of environmental release reservoirs are of great significance to their future ecological risks.

6.5.2 The Environmental Cumulative Estimation Method for Environmental Releasers

Similar to Sect. 6.4, we can obtain the total amount of releasers in a specific period by analyzing the annual material flow in a given country or region and accumulating the amount of environmental releasers during the study period based on acquiring the releasers each year. It will require a significant amount of effort to use this estimation method to estimate the total amount of environmental releasers during a given period. Therefore, we can establish a quantitative relationship between the amount of environmental releasers during a statistical period and several available statistics according to the material flow framework and the introduction of some parameters. However, how do we specifically establish the quantitative relationship? This question serves as homework for the students to consider and explore after class. Additionally, we recommend reading a number of documents for reference. Similar to Sect. 6.4, this method of estimating the amount of environmental releasers by means of material flow analysis can also be referred as the top-down approach for estimating environmental releasers.

In addition, students should consider, relative to the top-down approach, if there is a bottom-up approach for estimating environmental releasers. How would it be applied? What are the advantages and disadvantages? How can it be simplified? These questions serve as the scientific research issues for students to expand their thinking; textbooks will not provide this knowledge.

6.6 What Are the Uses of Material Flow Analysis?

With material flow analysis, we can acquire a significant amount of quantitative information, which is important for effective management of resources, the environment, and human socio-economic systems. In addition to establishing management indicators in Chap. 5, we provide representative examples of their application for student learning.

Application Example 1: Understand the relationship between human development and environmental change.

The quantitative relationship of parameters including resource consumption, pollution emissions, and social services in different regions can be obtained through material flow analysis; however, social and economic development is different in different regions. Data from various regions at different levels of development during the same period can be collated to quantify the quantitative relationship between human development and environmental change, as described in Chap. 2.

For example, the relationship between lead consumption and GDP in various countries as well as the relationship between lead consumption and the human development index are obtained by analyzing the lead flow in 52 countries around the world and selecting the GDP and human development index in various countries as indicators of human development, which verifies that lead consumption is growing with human development. (See recommended reading, Mao et al. [8].)

There are many uses for material flow analysis. Students should refer to the recommended reading and consider and discuss possible uses in classroom practice. (See class exercise I in this chapter.)

Classroom Discussions and Assignments

Exercises in Class: Several Applications of Logistics Analysis

Since the development of material flow analysis, the analysis methods for different substances have been developed and are commonly used in a variety of different areas. Students can consider and discuss the research in Table 6.2 or list studies of interest to them and record their possible application in table.

Comparison of Two Logistics Analysis Methods

We explored two methods for analyzing the flow of material, specifically, the tracking and fixed-point methods. Please compare the differences and advantages and disadvantages of these two methods. The results of the discussion should be recorded in Table 6.3.

Table 6.2 Several applications of logistics analysis

S/N	Research content	Application
1	The logistics comparison of the same material and the same region during different historical periods	
2	The logistics comparison of the same material in different regions	
3	A study of the correlation of different materials during the same period	
4	A study of the spatial distribution of a substance	
5	Analysis of the historical evolution of a substance	
6	Other studies	

Table 6.3 Comparison and analysis of two logistics analysis methods

Item	Common factors	Differences	
		Logistics tracking method	Logistics fixed-point method

Classroom Issues

Students choose a familiar area, discuss how to apply a logistics analysis method, determine how to solve the problem, and list the research outline for its typical resources or environmental issues.

Group Assignments

Based on selected questions, students locate logistics analysis papers that identify solutions to problems to share them in the form of classroom reports. The requirements are the same as the group assignment in Chap. 5.

Recommended and Referenced Literatures

1. Chen, W.Q., and T.E. Graedel. 2015. Improved alternatives for estimating in-use material stocks. *Environmental Science and Technology* 49 (5): 3048–3055.
2. Harper, E.M., G. Kavlak, and L. Burmeister. 2015. Criticality of the geological zinctin and lead family. *Journal of Industrial Ecology* 19 (4): 628–644.
3. Kavlak, G., and T.E. Graedel. 2013. Global anthropogenic tellurium cycles for 1940–2010. *Resources, Conservation and Recycling* 76: 21–26.
4. Liang, J., and J.S. Mao. 2014. A dynamic analysis of environmental losses from anthropogenic lead flow and their accumulation in China. *Transactions of Nonferrous Metals Society of China* 24 (4): 1125–1133.
5. Liang, J., and J.S. Mao. 2015. Source analysis of global anthropogenic lead emissions their quantities and species. *Environmental Science and Pollution Research* 22 (9): 7129–7138.
6. Mao, J.S., J. Cao, and T.E. Graedel. 2009. Losses to the environment from the multilevel cycle of anthropogenic lead. *Environmental Pollution* 157: 2670–2677.
7. Mao, J.S., J. Dong, and T.E. Graedel. 2008. The multilevel cycle of anthropogenic lead I: Methodology. *Resource Conservation and Recycling* 52: 1058–1064.
8. Mao, J.S., J. Dong, and T.E. Graedel. 2008. The multilevel cycle of anthropogenic lead II: Results and discussion. *Resource Conservation and Recycling* 52: 1050–1057.
9. Mao, J.S., and T.E. Graedel. 2009. Lead in-use stock: A dynamic analysis. *Journal of Industrial Ecology* 13 (1): 112–126.
10. Ohno, H., P. Nuss, and W.Q. Chen. 2016. Deriving the metal and alloy networks of modern technology. *Environmental Science and Technology* 50 (7): 4082–4090.

11. Reck, B., and T.E. Graedel. 2012. Challenges in metal recycling. *Science* 337 (6095): 690–695.
12. Lan, Ma., and Mao Jiansu. 2014. Quantitative analysis on the change in anthropogenic lead flows of China. *Environmental Science* 35 (7): 2829–2833.
13. Lan, Ma., and Mao Jiansu. 2014. The reasons for the changes in anthropogenic lead flows of China. *Environmental Sciences* 35 (8): 3219–3224.

7 Flow of Energy

7.1 Presentation of Core Issues

As previously mentioned, the goal of an anthropogenic flow of material is to meet specific needs of human beings. The transformation of material from natural resource into a product with specific service functions must undergo changes from multiple aspects, such as quantity, form, and space. It can be observed from physics and chemistry that these changes to the material must be implemented via an external force and with the use of energy instead of the material's own natural transformation process. Therefore, the anthropogenic flow of energy certainly exists within the material flow process of the socio-economic system of human beings. The flow of energy is of equal importance to the material flow. Additionally, how does energy flow? This is the topic to be discussed in this chapter.

To facilitate understanding, first, we will analyze the flow of energy based on the similar concept of material flow and then discuss how energy flows in a human society economy and how to evaluate it by learning from the natural flow process of energy. Finally, we will discuss energy consumption in our country through case analysis and introduce topics for further discussion.

7.2 What Is the Flow of Energy?

The movement of material cannot be separated from the force of energy. When material is transferred or exchanged in different systems, there is a flow of energy between a system and the external environment or between each component of the system. For example, solar energy is assimilated through plant photosynthesis, so that solar energy is transformed and absorbed by plants. When animals digest plants, the energy of the plants is partially transformed into the animals' bodies; thus, the transfer process of energy shows the flow of energy between different living organisms, which is called the flow of energy.

J. Mao et al., *Circular Economy and Sustainable Development Enterprises*,
https://doi.org/10.1007/978-981-10-8524-6_7

Based on the principle of whether there is interference from human activity, the flow of material has been divided into the natural flow and the anthropogenic flow of material by human beings. Similarly, the flow of energy is also divided into natural and anthropogenic flows. The natural flow of energy refers to the flow of energy in a natural ecosystem, while the anthropogenic flow of energy refers to the flow of energy in a socio-economic system. An anthropogenic flow of energy is formed via a series of processing and transformations of natural resources to meet human beings' demands in many areas, such as living, life, and production. It is not difficult to imagine that the anthropogenic flow of energy can be treated as a special segment of the natural flow of energy and the portion of the flow utilized and interfered with by human beings in the natural flow of energy will be restricted by the natural flow of energy as well.

7.3 What Is the Natural Flow of Energy?

7.3.1 Natural Flow of Energy

To facilitate understanding the flow of energy, first, review the description of the natural flow of energy in ecology.

Energy is the basis of all biogenic activities in an ecosystem and provides the power to sustain the ecosystem. There are three basic processes in the energy flow of an ecosystem, as described below.

(1) Fixation process of energy—Inorganic matter is transformed into organic matter and solar energy is transformed into biochemical energy for storage nearly simultaneously via photosynthesis of green plants (e.g., plant, algae, and some bacteria). These green plants are called the producers in the ecosystem food chain.
(2) Decomposition-reduction process—Organic matter is deacidized into water, CO_2, and other inorganic matter primarily via a decomposer; for example, the inorganic matter in animal/plant remains is decomposed into simple inorganic matter via the saprophytic microbe in the ecosystem and becomes nutrients for green plants.
(3) Storage and mineralization process—After deceased living bodies are buried, the organisms become part of the mineralization process to form fuel minerals. For example, coal is a sedimentary mineral substance transformed via the endless (generally, from tens to hundreds of millions of years) biochemical action and physical–chemical reaction of plant remains.

In general, the natural flow of energy can be regarded as a process, for example the photosynthesis of green plants. Solar energy enters the ecosystem and flows through each trophic level along with the food chain and finally transforms into a surface mineral fuel via the mineralization of organisms.

7.3.2 Natural Flow Characteristics of Energy

Compared with the natural flow of material, the natural flow of energy has the typical characteristics described below.

(1) Unidirectional—The energy in an ecosystem follows a one-way flow along with the trophic level from low to high. Put differently, solar energy enters an ecosystem once and flows through each trophic level instead of returning to the original organism. This is entirely different from the circular flow of returning to a specific ecological subsystem in the natural flow of material.
(2) Degressive—When energy flows through each trophic level of an ecosystem, its quantity decreases at each level, forming the pyramid of energy, as shown in Fig. 7.1. While the quantity of material is constant either in a natural or anthropogenic flow, it is distributed into different system components.
(3) Gradual improvement of energy quality—When energy flows into a high trophic level from a low trophic level in an ecosystem, the quality of energy gradually improves, showing that the content in a unit of material improves gradually. For example, 1 kg of vegetables contains approximately 700 kJ of energy, while 1 kg of meat contains approximately 6000 kJ of energy. In the natural flow of material, only the service object changes, but the service quality in the entire natural system retains the same important role.
(4) Quantity dynamics of energy—For a specific ecosystem, the quantity and quality of energy constantly vary based on the change of energy input/output, components of the internal system and biological consumption. For the

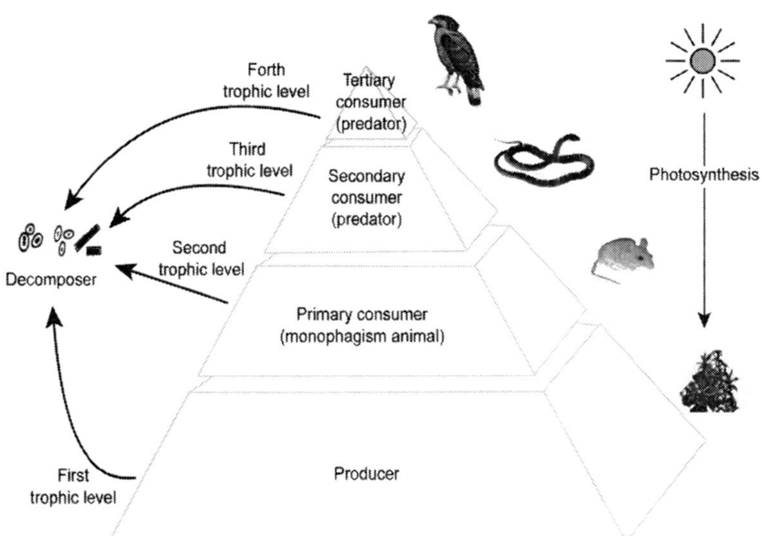

Fig. 7.1 Pyramid of energy. *Source* Enger and Smith [8]

material flow in a thermodynamic system, because the system consists of specific material particles and the quantity of material follows the law of conservation of mass, the gross amount of material in the system is constant.

7.3.3 Evaluation of the Natural Energy Flow—Energy Conversion Efficiency

In ecology, ecological efficiencies are used to evaluate the validity of energy flow in an ecosystem. Ecological efficiency refers to the efficiency of the transfer and transformation of the energy parameter at each trophic level and in the interior. Generally, four parameters of intake (*I*), assimilation (*A*), respiratory volume (*R*), and production (*P*) are used to measure the ecological efficiency of the energy flow in an ecosystem. Intake (*I*) represents the energy used by an organism; plants absorb light energy and animals digest food. Assimilation (*A*) is the food energy absorbed from the food consumed by the organism and in plants it is the energy provided by photosynthesis. In animals, food energy is absorbed inside the alimentary canal, and for a decomposer, food energy is absorbed from the extracellular product. Respiratory volume (*R*) refers to the total energy consumed by an organism during metabolism and other activities, and production (*P*) refers to the remaining net assimilated energy after consumption by the organism. In plants, it is the net primary production (*NP*), and in animals, it is the energy that is assimilated minus the respiratory volume. Ecological efficiency between trophic levels is used to measure the capacity of the transformation efficiency and energy flow pathway between trophic levels; a significant amount of the assimilation energy can be revealed by the internal ecological efficiency of each trophic level. Figure 7.2 shows the change process of energy in an ecosystem.

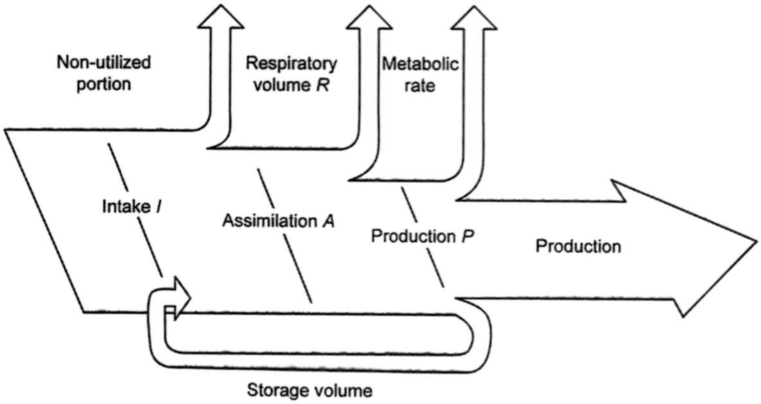

Fig. 7.2 Sketch map of energy transformation in ecosystem

In an ecosystem, the transformation efficiency of energy from one trophic level to another is in the range of 5–30%, varying between the different trophic levels. Generally speaking, the transformation efficiency of energy from plants to herbivores is approximately 10% and the transformation efficiency from herbivores to predators is approximately 15%.

The law of thermodynamics can clearly explain the transfer and transformation of energy in an ecosystem and can determine the limits of using energy in an ecosystem, the food chain links and the quantity of organisms in each trophic level.

7.4 What Is the Anthropogenic Flow of Energy?

7.4.1 Concept of the Anthropogenic Flow of Energy

The anthropogenic flow of energy is related to the natural flow of energy. It refers to the entire process used to meet human beings' needs in production and life. Energy is transformed from primary energy into a form capable of being used by human beings directly and delivered to the ultimate user after processing, transformation, transportation, and distribution. The set of processes that energy experiences in this flow is called the anthropogenic flow of energy, as shown in Fig. 7.3.

The first box in Fig. 7.3 refers to the process of extracting the natural resources from the natural system, such as coal or petroleum mining. These substances are capable of being used as energy for human beings and can be called primary energy. In the second box in the figure, a human activity transforms the primary energy into another form of energy available for use by human beings. Based on human production and transformation, this is called secondary energy; e.g., by means of coal-fired power generation, part of the energy in the coal is transformed into electric energy. The third box distributes the secondary energy for use, for example, by means of a high voltage grid, the electric energy generated in the Shanxi coal-fired power plant is transmitted and distributed to the Beijing area and then transported into each enterprise's manufacturing shop or residential user via the Beijing network system. Thus, the terminal equipment for the electric energy is the production equipment in each workshop and each piece of electrical equipment. The fourth box further transforms the secondary energy into the energy demanded by users by using terminal equipment; for example, electric energy is transformed into heat energy to cook food via a user's electric cooker and electric energy is

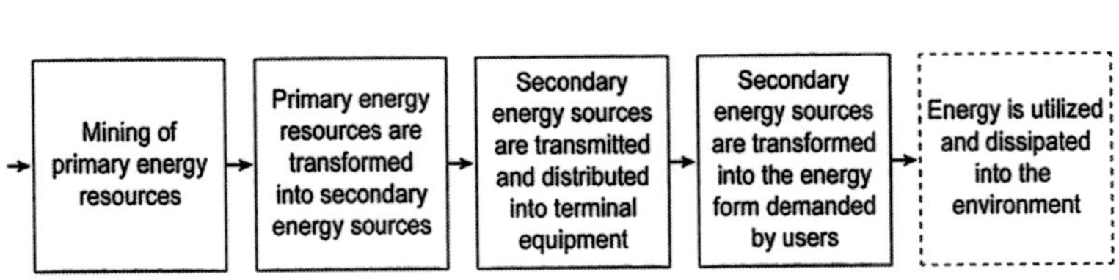

Fig. 7.3 Sketch map of the anthropogenic flow of energy

transformed into luminous energy to light rooms via electric lamps. The final box is the final use procedure. Energy drives materials to a corresponding transformation process, and conversely, energy constantly dissipates into the environment; for example, during cool seasons, the quantity of heat carried by the water in a radiator will continually dissipate into a room.

It is not difficult to imagine that the natural flow of energy occurs in the natural system beyond human beings to serve the natural ecosystem. Additionally, the anthropogenic flow of energy occurs in the socio-economic system to meet human beings' demands both for production and living. There is a significant difference between the anthropogenic flow of energy and its natural flow.

Due to the diversity of human beings' demands and the population distribution, the anthropogenic flow of energy is diversified, regional and complex; it is the result of the combined action of human activities and natural factors.

In addition, the following two points deserve attention: one is the difference between "energy source" and "energy." An energy source is the source of the energy and energy represents the quantity of "energy." The other refers to "energy source" in daily life, which also refers to the source of energy in the terminal equipment, e.g., "power source" usually refers to an electrical socket. Please ensure to distinguish these terms in use.

7.4.2 Anthropogenic Flow Process of Energy

(I) Framework for the anthropogenic flow of energy

If the energy source capable of being directly used by human beings is regarded as the service provided by the energy system, it is available to build a model framework of the energy flow in the energy service system based on the method for analyzing material flow in the product system.

In the product system, the flow process of material primarily includes the basic stages of mining the natural resources, processing the materials and manufacturing, using, scrapping, recovering and reusing the product. Each basic production chain requires the site transfer of the product or its input materials, specifically, the transportation process. The service provided by the product system for human society is represented by a product with a specific function, such as an automobile carrying passengers or houses for human habitation. It can also be expressed by the quantity of the services enjoyed by human beings, such as a trip from Beijing to Hong Kong by ten travelers and the residence where a family has lived for 30 years.

Similarly, in this special energy "product" system, several life cycle periods are required as follows. First is the extraction of primary energy, e.g., coal mining and oil exploitation, and second is energy conversion, the transformation from primary energy to secondary energy, e.g., petroleum is refined into gasoline to be the combustion power of the motor vehicle, coal is burnt into heating power as the heat energy for heating in winter. Next, the secondary energy must be delivered to terminal equipment arranged in the users' locations via an energy transportation

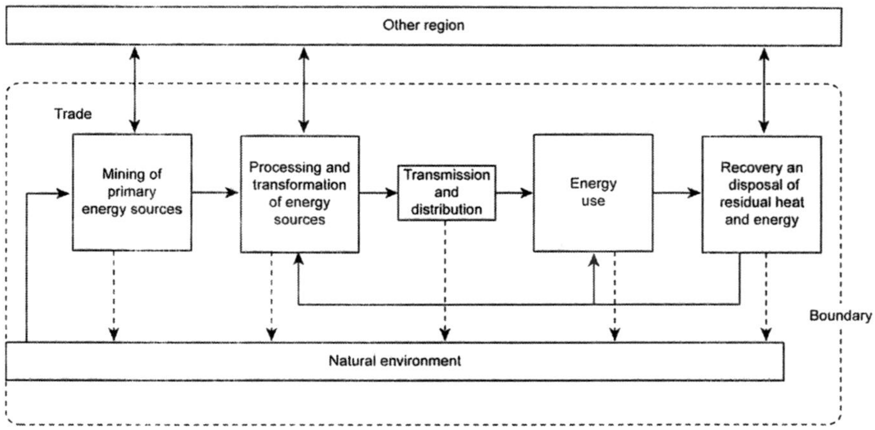

Fig. 7.4 Frame model of the anthropogenic flow of energy

network or energy transportation equipment. Powered by secondary energy, the secondary energy is transformed into a form of energy that satisfies the special needs of human beings via terminal equipment; e.g., by using electricity, an air conditioner offers users air cooling and lighting equipment is used to provide light. Thus, the energy product system satisfies its purpose of providing services for human beings. Energy also follows the flow process from primary energy in natural resources to available energy for special services for human beings, completing the flow of the energy source in the social economy. Finally, energy returns to the natural environment via the dissipation process. The entire anthropogenic flow of energy process is shown in Fig. 7.4.

It can be observed that the anthropogenic flow of energy primarily includes the mining of primary energy, e.g., the mining of coal and petroleum; energy processing and transformation, e.g., an energy source is transformed into steam via coal combustion and then a turbine is driven by the steam to generate electricity to fuel electric products; and transmission and distribution of energy, e.g., electricity generated in a thermal power plant is transported to the end user via an electric system. In addition, the trade of primary energy and an energy product exist, e.g., international trade and regional trade of electric power. A portion of the residual heat and energy are recycled or reprocessed for use. Finally, a portion of the energy in each phase dissipates into the natural environment.

From the anthropogenic flow of energy, it is not difficult to observe that by recycling residual heat and energy there is a circular energy flow, namely, energy returns to the phase of energy processing and transformation, e.g., the backwater of a heating system maintains a specific temperature (e.g., 70 °C) and retains a specific amount of heat energy that returns to the boiler room with the backwater circulation; the carried heat energy also returns to the energy processing and transformation phase to form a circular flow of energy. This is another significant difference between the anthropogenic flow of energy and its natural flow.

(II) Energy sources and examples
1. Energy sources and classification

In the anthropogenic flow of energy, we are concerned with the portion of energy capable of meeting human production processing or living needs. Therefore, the source of energy refers to the material resource capable of directly providing or being transformed to provide energy and power for human production and life.

For the convenience of management and application, the source of energy is usually classified; Table 7.1 shows several types of energy sources under different classification methods.

An energy source is a source of energy. And conversely, the direction of the energy flow is called the collection of energy and refers to the energy lost and discharged by human beings into the area beyond the socio-economic system. After entering the environmental system, the quality of this portion of energy is very low; therefore, it is no longer able to serve human beings.

2. Several examples of conventional sources of energy

(1) Fossil fuel—This primarily includes coal, petroleum and natural gas. They were transformed from the fossils of animal and plant remains millions of years ago after the long-time action of soil and rock pressure and internal heat of the earth and are in the forms of solid, liquid and gas. Their formation requires millions of years or more; therefore, relative to the utilization of energy for human beings, they are a non-regenerated energy source. Fossil fuel is the primary type of energy utilized by human beings to date. The usage of coal has been maximized especially, as one-third of the global energy has been utilized, and nearly 75% has been used in our country. Because fossil fuel releases a variety of air pollutants during the combustion process, in recent years, multiple cleaning measures have been employed to reduce the degree of contamination, and concurrently, new clean energy has been discovered.

(2) New source of energy—This generally refers to energy sources beyond traditional energy. They are the renewable energy sources developed and utilized via new techniques. It primarily includes solar, wind, biomass, tidal, geothermal and nuclear energy. Of these, solar energy is produced via the continuous nuclear fusion reactions of the solar interior. Although the energy radiated from the sun to the earth's atmosphere is only 1/2,200,000,000 of the total radiation energy (approximately 3.75×1026 W), it is nearly 173,000 TW. Put differently, the energy of the sun irradiating on the earth each second is equivalent to 5 million tons of coal. The majority of the wind and biomass energy are from the sun; therefore, solar energy is the richest primary energy and should be greatly developed and utilized. Nuclear energy is the energy released from an atomic nucleus via a nuclear reaction, and many nuclear power plants have been built around the world to transform the energy released by the

Table 7.1 Examples of the common classification methods of energy sources

Classification standard	Broad heading	Explanation	Major energy sources
Based on the source of energy	Energy of celestial bodies beyond the earth		Solar radiation energy
	Energy contained in the earth itself		Geothermal energy and atomic energy
	Energy produced by the interaction of the earth and other celestial bodies		Tidal energy
Based on whether there is human being processing and transformation	Primary energy	Energy capable of being directly taken from nature without any change or transformation	Crude oil, raw coal, natural gas, biomass energy, water power, nuclear fuel, solar energy, geothermal energy, tidal energy
	Secondary energy	Energy products obtained via the processing or transformation of the primary energy	Electric power, steam, coal gas, gasoline, diesel oil, heavy oil, liquefied petroleum gas, alcohol, marsh gas, hydrogen, coke
	End energy	The energy capable of being utilized by consumers via energy-using equipment	Electric power to drive electrical equipment, quantity of heat or refrigeration capacity provided by air-conditioning equipment
Based on use technology	Conventional source of energy	Energy that has been produced on a large scale and extensively used under existing economic and technological conditions	Coal, petroleum, natural gas, hydraulic energy, nuclear fission energy
	New source of energy	Related to a conventional source of energy and developed and utilized via new techniques	Solar energy, ocean energy, geothermal energy, biomass energy

(continued)

Table 7.1 (continued)

Classification standard	Broad heading	Explanation	Major energy sources
Based on whether there is commodity circulation	Commercial energy	Commercial energy is the energy consumed via commodity circulation, e.g., market trade, management, and allocation of government energy	Currently, it primarily includes coal, petroleum, natural gas, water, electricity, and nuclear power
	Non-commercial energy	Non-commercial energy refers to the energy from core wood and crop residuals (e.g., straw) which can be used on the spot. These sources of energy are renewable	Core wood, crop residual (e.g., straw)
Based on whether environmental pollution is caused	Clean energy		Hydraulic power, electric power, solar energy, wind energy, nuclear energy
	Non-clean energy		Coal, petroleum

fission reaction into electric energy. It is reported that 80% of the electric energy production in France is currently from nuclear power. In our country, the nuclear generating capacity provides 2.3% of the total power generation. Wind energy is the energy carried in air flow movement. Geothermal energy is the energy released by the disintegration of molten magma and radioactive substances in the bowels of the earth.

(III) Mining of energy sources

Mining of energy sources refers to the process of extracting the energy-inclusive substance from the natural system; it is the origin of the anthropogenic flow of energy. The energy directly obtained from the exploitation process of the energy is called the primary energy. For example, for the coal mine found via exploration, surface soil and rock on the surface of the coal bed are removed to expose the coal bed for strip mining. Alternatively, if the coal bed is too deep, the coal is excavated from underground by means of a tunnel from the surface to the coal bed. A similar process is used in the mining of petroleum and natural gas. This is the process of obtaining primary energy products via a direct action between human beings and natural resources. Primary energy can be regarded as the stage product in this process, such as raw coal or crude oil. The majority of new energy sources have rich reserves; thus, they should be developed technologically and fully utilized. For the majority of conventional energy (e.g., mineral fuel), due to limited resources, mining might cause resource shortages or depletion; thus, resource development and protection should be implemented simultaneously. To understand the exploitation of energy sources more clearly, it is recommended students watch the short film *Revelation of Civilization: Energy Exploitation on the Earth* via Phoenix TV, on the following Web site:

Revelation of Civilization: Energy Exploitation on the Earth. Phoenix TV, 2012-05-17. http://v.ifeng.com/news/society/201205/b1ac9458-3992-4868-926f-66e0c3a8ec73.shtml.

(IV) Transformation of energy sources

Although the sources of energy exist in the natural world in multiple forms, it is difficult for them to meet human needs directly. For example, sunlight is able to provide room illumination by day, but it is unable to meet lighting needs at night. Therefore, energy in the natural world must be transformed and reserved.

The transformation of energy sources is the process of transforming primary energy into forms of energy capable of being directly utilized by human beings; the transformed energy is called secondary energy. For example, coal and petroleum are primary energy; the electric energy obtained from coal-fired power generation and gasoline or diesel oil formed via petroleum refining are secondary energies. The technique used in an energy transformation process is called an energy transformation technique; for example, heat energy of sunlight is collected via a condensation panel to drive turbine rotation, transforming heat energy into electric energy. In hydroelectric generation, the potential energy of water spins a water turbine, and a generator driven by the turbine transforms the energy into electric energy. Coal transforms into heat energy via combustion, the heat energy produces steam to spin a turbine creating mechanical energy, and then, an electric generator is driven to transform the mechanical energy into electric energy. Compared to primary energy, transformed secondary energy has higher end utilization efficiency and is more convenient and clean to use.

During the transformation process of energy sources, there is an inevitable conversion loss of energy. For example, a specific amount of coal is used to generate electricity. A portion of the energy of the coal remains in the non-burnt coal granules, a portion is dissipated via the chimney in the form of heat or becomes the radiant heat of the boiler or steam pipeline for emission, a portion is dissipated in the form of waste heat, a portion is frictional loss, and the remainder is transformed into electric energy. Although technology is constantly being improved and equipment is continuously being perfected to improve energy conversion efficiency, it is difficult to avoid a portion of the energy dissipating into the environment during such a process. Generally, the energy conversion ratio in the power generation of a turbine is approximately 30%.

Efficiency of energy processing and conversion is defined in the *China Statistical Yearbook*, which refers to the ratio between the quantity of the various energy products after processing and transforming the energy sources within a certain period, and the quantity of the energy sources input for processing and transformation within the same period. It is reported that the total energy conversion efficiency in our country improves constantly and is currently at 73%, which is up from 70% at the end of the twentieth century. Additionally, the energy conversion efficiency from power generation and power station heating has also increased to approximately 43% from 36%, and the energy conversion efficiency is higher in the coking and oil refining process, with values of 95 and 97%, respectively, reflecting an increase of 1–2% within the same period. This indicator is an important parameter to observe the degree of advancement and management level of the energy processing and transformation devices and manufacturing techniques. The indicator can be calculated by the following equation:

$$\text{Processing and transforming efficiency of energy sources} = \frac{\text{Processing and transforming output of energy sources}}{\text{Processing and transforming input of energy sources}} \times 100\% \quad (7.1)$$

In energy transformation, energy should be transformed into a form capable of being directly utilized by end users. The security and suitability of energy sources in transmission and distribution must be considered, and they should be transformed into a form that is easy to store, carry, and transport.

The secondary energy formed through production and transformation is the energy source for human production equipment or living facilities.

(V) Transmission and distribution of energy sources

Transmission and distribution of energy sources is the process to transport and allocate energy, and it is also the process to deliver the energy to the terminal energy-using equipment with the quality and quantity required by the final users. It includes the transfer from the site of energy production to the site of energy utilization and the redistribution of the energy to the end users.

Delivery of energy refers to the change of region and is primarily completed via an energy delivery system; e.g., heat energy is delivered to heating equipment in buildings via a heating supply system and a regional heating system. Electric energy is delivered to specific electric equipment from a power plant via a power supply system, and currently, our country has built multiple energy transportation systems, including the famous "West-East Natural Gas Transmission" gas supply system, common gas supply systems for homes, and power supply systems. These energy supply systems are completed primarily via a pipe network. In addition, there are many other means of transportation; for example, vehicles are used to complete the regional transfer of energy, petroleum products are delivered to different places via an oil tank truck, and a gas tank is delivered to each scattered living user via automobile.

Distribution of energy sources is the redistribution of its quantity. Based on quality requirements, energy sources are allocated to the final users, including the quality and quantity of energy sources. The following are examples of quality requirements: The steam demanded by different production processes should have different pressures, and the voltage demanded for different areas or electricity utilization locations should be different (i.e., commercial power is 380 V, and civil electricity is 220 V). Regarding quantity, through an energy distribution device, e.g., a switchboard, electric energy of a certain circuit is allocated to a nearby load; concurrently, the load is protected, monitored, and controlled to ensure the normal conditions of various pieces of terminal equipment. For example, 0.56 kW h/d of electric energy is required by a Haier 216 L refrigerator under normal operations, while the power to maintain the continuous running of a specific air conditioner requires approximately 3.5 kW of electric energy.

A process to convert the variability of the energy flow intensity in the natural world to the stability of the energy demanded by end users must be implemented by means of energy source reserves. Usually, energy storage devices include items such as oil storage tanks, gas storage tanks, and cells.

Similarly, in the transmission and distribution of energy, there is also a loss of energy caused by the temperature difference inside and outside of the system and the (thermal) electrical conductivity of the transmission–distribution pipeline materials. Based on data in the *China Statistical Yearbook*, it can be estimated that transport efficiency of different energy delivery systems in our country ranges from 93 to 98%. With the improvement of the transport efficiency via technical progress within the past 50 years, the loss of energy is reducing constantly; e.g., the grid loss today is approximately 6% compared to 8% in the 1980s.

(VI) End use of energy sources

The end use of energy sources refers to the process of the end user using energy. The word "end" refers to energy sources that are not processed or transformed beyond terminal equipment; there is no additional trade or commercial activity. Usually, the end use of energy sources includes plant and life uses. In production, energy sources are primarily used to power equipment to complete the transformation of materials via specific manufacturing techniques, and in life, energy sources are primarily used to satisfy illumination needs and operation of various household appliances. In the end use of energy, its form may be transformed; e.g., in the operation of a refrigerator, the compressor is driven by electric energy to power the refrigeration system, achieving heat transfer and showing that heat energy is input to obtain a refrigeration effect (a type of heat energy), and electric energy is transformed into light to brighten rooms via lighting fixtures. By means of terminal equipment, the energy sources achieve human beings' ultimate demands for heat energy and illumination.

It is not difficult to imagine that during final use, not all energy input into the terminal equipment can be totally transformed into the specific energy demanded by human beings. The work efficiency of terminal equipment is the percentage of energy demanded by human beings of the total energy input into the terminal equipment. The work efficiency of terminal equipment varies among different equipment. For example, the efficiency of lamps and lanterns is defined as the ratio between the luminous flux value they emit (measured under specified conditions) and the sum of the measured luminous flux values emitted by all light sources in lamps and lanterns. The efficiency varies between different types of lamps and lanterns. For example, the efficiency of project and fluorescent lamps is generally in the range of 40–55%, and the efficiency of an LED lamp can reach 85–90%.

In summary, the anthropogenic flow of energy is the flow of energy in a socio-economic system, beginning with natural resources. It consists of the mining, processing and transformation, transmission and distribution of human primary energy, and use of the end energy to meet the entire process of human production and living needs. The form of energy that exists at the end point of energy's anthropogenic flow can be regarded as the "product" of energy's anthropogenic flow system, and it is in a form that can meet human demands.

7.4.3 Evaluation of the Anthropogenic Flow of Energy—Energy Efficiency

In the anthropogenic flow of energy, energy efficiency can be defined as the ratio between the available energy obtained by human beings and the primary energy input into system initially. It is expressed as follows:

$$e_{\mathrm{N}} = \frac{E_{\mathrm{valid}}}{E_{\mathrm{primary}}} \times 100\% \tag{7.2}$$

where e_{N} represents energy efficiency, the subscript N is the second letter in energy and is used to distinguishing this parameter from the ecological efficiency discussed previously, E_{valid} is the available energy used by the end users, and E_{primary} is the quantity of energy sources taken from the natural system.

If energy flows through each phase in series as the model of energy anthropogenic flow shown in Fig. 7.4, and there is a certain energy conversion or utilization efficiency in each phase, the serial number of each phase is represented by subscript i, the total energy efficiency of the entire anthropogenic flow process of energy would be the product of energy efficiency of each phase, expressed as follows:

$$e_{\mathrm{N}} = \prod_{i=1}^{n} e_{\mathrm{N}i} \tag{7.3}$$

where $e_{\mathrm{N}i}$ is the energy efficiency of the ith phase and n is the sum of the phases in the system.

Given the transformation efficiency of primary energy is 50%, there is a 10% loss in transmission and distribution, and the end energy efficiency is 60%, the total energy efficiency would be only 27%. This means that 1 kJ of energy is obtained from natural resources and only 300 J is used to provide final service to human beings.

7.5 Case Analysis: What Is the Energy Consumption Situation in Our Country?

7.5.1 Structure of Energy Consumption

Energy consumption refers to human beings' use of energy sources to meet their production and living needs. The consumed source of energy is also called the amount of energy consumption. There are different modes of energy consumption in different fields. If energy consumption is classified based on the energy consumption department, we can estimate the amount of energy consumed and its ratio within each statistical period (e.g., month, quarter, and year) for a country or region. The indicator can reflect the structural features of energy consumption of a country or region. Figure 7.5 is the structure of energy consumption classified based on the energy consumption department in 2015. It can be observed in Fig. 7.5 that industry is the most significant energy consumer in our country.

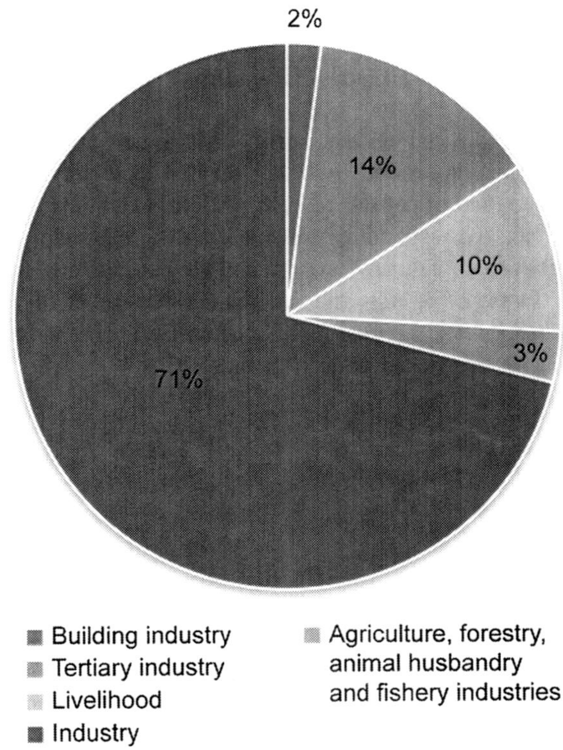

Fig. 7.5 Composition of energy consumption in China 2015

7.5.2 What Is the Energy Efficiency of Industrial Sectors in Our Country?

Because industry is the energy consumer in our country, its level of its energy efficiency has an important effect on our country's energy efficiency. Therefore, beginning with industry, the analysis of energy efficiency in the industrial sector will provide an important information base for energy conservation in China.

(I) Composition of industry sectors

It is stipulated by China's National Bureau of Statistics that industry is the generic term for material production departments engaged in the mining of natural resources, processing and re-processing of excavated products and agricultural products. According to China's National Industries Classification standards, industry further includes three major types of mining industries, manufacturing industries and electric power, heating power, fuel gas, water production and supply industries. It can be further subdivided into 39 sectors. To facilitate the settlement of energy efficiency of each industry sector, the name of each industry is represented by an English abbreviation, as shown in Table 7.2.

(II) Model of an industrial system

In an industrial system, energy will be consumed in the operation of any industry and specific services will be provided to society. For the entire industrial system, the amount of consumed energy sources and the provided social services is the sum of the corresponding values constituting industrial sectors. The relationship between an industrial system and the external environment can be seen in Fig. 7.6.

In Fig. 7.6, subscript i is the serial number of a specific industrial sector and subscript n is the total number of industries in the industrial system. The following relationships exist in each parameter:

$$E = \sum_{i=1}^{n} E_i \tag{7.4}$$

$$G = \sum_{i=1}^{n} G_i \tag{7.5}$$

where E is the input of energy sources in an industrial system and is represented by energy consumption in units of tce/a and G represents the system services offered to society and is represented by economic value added in units of ten thousand Yuan/a.

Table 7.2 Names of industry sectors and their codes

Industry category	GB/T 4754—2002		GB/T 4754—2011	
	Name of industry	Code	Name of industry	Code
Mining industry	Coal mining and washing industry	CMW	(Same)	CMW
	Petroleum and natural gas mining	PGX	(Same)	PGX
	Ferrous metals ore mining and selection industry	FMM	(Same)	FMM
	Nonferrous metals ore mining and selection industry	NFM	(Same)	NFM
	Nonmetal ores mining and selection industry	NOM	(Same)	NOM
			Mining support activity	AMA
	Other mining industry	OOM	(Same)	OOM
Manufacturing industry	Agricultural subsidiary foodstuff processing industry	AFP	(Same)	AFP
	Food manufacturing	FOM	(Same)	FOM
	Beverage manufacturing	BEM	Liquor, drink, and purified tea manufacturing industry	BEM
	Tobacco industry	TOM	(Same)	TOM
	Textile industry	TXM	(Same)	TXM
	Textile and garment, shoes and cap manufacturing industry	TWM	Textile and garment and fashion industry	TWM
	Leather, fur, feather (down), and their products industry	LFM	Leather, fur, feather, their products and shoemaking industry	LFM
	Wood processing, wood, bamboo, rattan, palm, and straw-made articles industry	WBP	(Same)	WBP
	Furniture manufacturing	FNM	(Same)	FNM
	Papermaking and paper products industry	PAM	(Same)	PAM
	Copy of printing industry and recording medium	RMP	Printing and recording medium copy industry	BMP
	Stationery and sporting goods manufacturing industry	ARM	Stationery, industrial art, sports and entertainment goods manufacturing industry	ARM
	Petroleum refining, coking, and nuclear fuel processing industry	FUP	(Same)	FUP
	Chemical materials and products manufacturing industry	CMM	(Same)	CMM
	Pharmaceutical industry	MEM	(Same)	MEM
	Chemical fibers manufacturing industry	CFM	(Same)	CFM
	Rubber products industry	RUM	Rubber and plastic products industry	RPM

(continued)

Table 7.2 (continued)

Industry category	GB/T 4754—2002		GB/T 4754—2011	
	Name of industry	Code	Name of industry	Code
	Plastic products industry	PLM	(Combined)	
	Nonmetallic minerals products industry	NMM	(Same)	NMM
	Ferrous metal smelting and rolling industry	FMS	(Same)	FMS
	Nonferrous metals smelting and rolling industry	NFS	(Same)	NFS
	Metal products industry	MPM	(Same)	MPM
	General equipment manufacturing industry	GMM	(Same)	GMM
	Special equipment manufacturing industry	SMM	(Same)	SMM
			Automobile industry	VMU
	Transport and communication facilities manufacturing industry	TRM	Manufacturing industry of railway, shipping, aerospace, and other transportation equipment	TRM
	Electric apparatus and equipment manufacturing industry	EEM	(Same)	EEM
	Manufacturing industry of communication device, computer, and other electronic equipment	CEM	Manufacturing industry of computer, communication, and other electronic equipment	CEM
	Manufacturing industry of instrument and meter, stationery, and office machinery	ICM	Instrument and meter manufacturing industry	ICM
	Artware and other manufacturing industry	AOM	Other manufacturing industry	AOM
	Waste resources and scrap materials recovery processing industry	WRD	Waste resources integrated utilization industry	WRD
			Metalware, machinery, and equipment repair and maintenance	MPR
Production and supply of electric and heating power and gas and water	Production and supply of electric power and heating power	EHP	Production and supply of electric power and heating power	EHP
	Gas production and supply	GPS	(Same)	GPS
	Water production and supply	WPS	(Same)	WPS

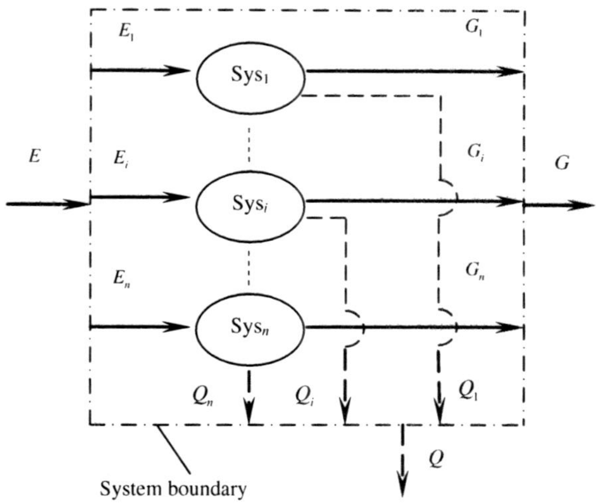

Fig. 7.6 Relationship frame between industrial system and its environment. Reprinted from Ref. [4], with permission from Environmental Science (Chinese)

(III) Industry energy efficiency

Based on the concept of ecological efficiency, an industrial system is selected as the research object, energy consumption and industrial added value of the industry (or sector) are selected as the environmental load and industrial output of the industry, respectively, and the industry energy efficiency can be represented as follows:

$$e_{\mathrm{N}} = \frac{G}{E} \tag{7.6}$$

where e_{N} is the energy efficiency in units of ten thousand Yuan/tce and the meanings of other symbols are the same as in previous equations.

It can be seen that the higher the energy efficiency, the less energy consumed to obtain the same value added.

For a specific industry *i*, Eq. (7.6) can be expressed as follows:

$$e_{\mathrm{N}i} = \frac{G_i}{E_i} \tag{7.7}$$

In China's statistical data, industrial added value and energy consumption of a specific region are the sum of the added value and energy consumption of each sector in the region; furthermore, for a specific industrial sector, there is a relationship [Eq. (7.7)] between its added value and energy consumption. Therefore, through settlement, the relationship between the energy efficiency of the industry and the energy efficiency of the industrial sector can be obtained as follows:

$$e_{\mathrm{N}} = \left[\sum_{i=1}^{n} (f_{G_i} \cdot e_{\mathrm{N}i}^{-1}) \right]^{-1} \tag{7.8}$$

where f_{Gi} is the contribution rate of the ith industry to the industrial added value as follows:

$$f_{G_i} = \frac{G_i}{G} \tag{7.9}$$

Furthermore,

$$\sum_{i=1}^{n} f_{G_i} = 1$$

It can be seen from Eq. (7.8) that industrial energy efficiency is related to its economic structure and the energy efficiency of each industry. To improve the energy efficiency of the entire industrial system, not only must the energy efficiency of each industry be improved, but the industry structure must also be adjusted to improve the contribution of the industry sector with higher energy efficiency to the economy. This conclusion is also suitable for other industrial ecological efficiency and is the important basis for adjusting the national economic structure.

(IV) Composition of industrial energy consumption

To clarify the composition of industrial energy consumption in China, energy consumption of a specific industrial sector is divided into the total industrial energy

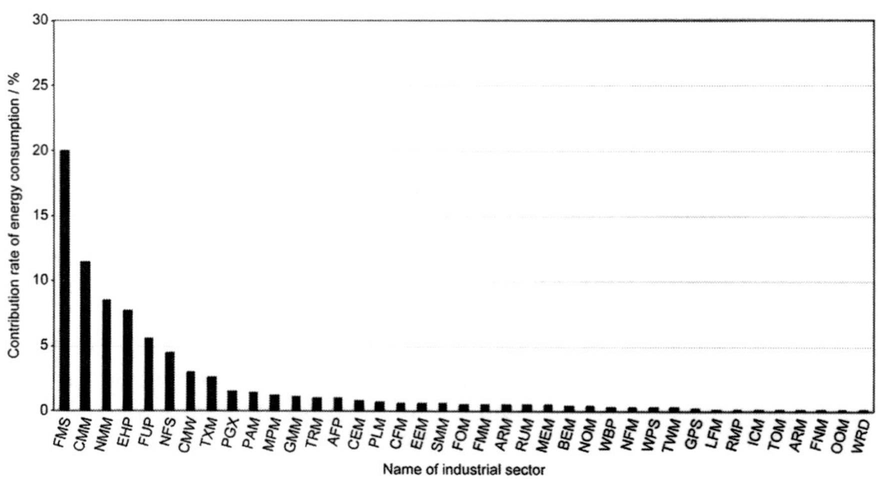

Fig. 7.7 Rank ordering for the contribution rate of industrial sectors to the industrial energy consumption in China. *Note* Based on GB/T 4754-2002. Reprinted from Ref. [4], with permission from Environmental Science (Chinese)

consumption within the same period to obtain the contribution rate of that industrial sector to the industrial energy consumption. The results of each industrial sector's contribution rate to the industrial energy consumption calculated from China's statistical data are sorted from high to low and presented in Fig. 7.7.

The code for each industrial sector in the figure is as shown in Table 7.2.

It can be observed from Fig. 7.7 that the five industrial sectors with maximum energy consumption in our country include the ferrous metal smelting and rolling industry (code FMS), the chemical materials and products manufacturing industry (code CMM), the non-metallic mineral products industry (code NMM), the production and supply of electric power and heating power industry (code EHP), and the petroleum processing, coking and nuclear fuel processing industry (code FUP). Of these, the ferrous metal smelting and rolling industry (code FMS) consumes the most energy, or approximately 20% of the total industrial energy consumption. The contribution rate of each other industrial sector is less than 10%, meaning that for industrial energy saving, the ferrous metal smelting and rolling industry should be the focus.

(V) Sector allocation of industrial energy efficiency

Based on the energy consumption of each industrial sector, the value of industrial output and other data provided by the China Statistical Yearbook and Eq. (7.7), the energy efficiency of each industrial sector within a specific statistical stage can be calculated. To observe the differences among different sectors from the perspective of energy efficiency more clearly, the values are ranked and presented in Fig. 7.8. It can be seen that the energy efficiency of the tobacco industry (code TOM) in 2007 was the highest, exceeding the next closest sectors by nearly double.

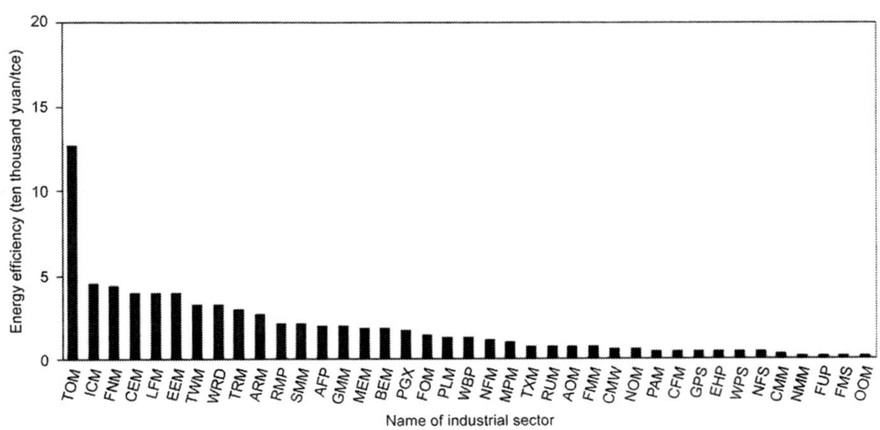

Fig. 7.8 Rank ordering of energy efficiency of industrial sector in China (2007). *Note* Based on GB/T 4754-2002. Reprinted from Ref. [4], with permission from Environmental Science (Chinese)

Based on similar methods, the energy efficiency in different areas and historical periods can be estimated, and the factors for low energy efficiency are identified as reference for structural adjustment and technology upgrade management.

Classroom Discussions and Assignments

Topic Discussion

Because we understand the anthropogenic flow process of energy, we must improve the flow. The analysis of energy flow in specific areas is an important basis to identify and improve the energy flow. Therefore, each group is requested to choose a specific area and discuss how to improve its utilization of energy sources. Please discuss and list the research outline of the analysis of energy flow in the special scale.

Example of the Topic Screening Process

The utilization of energy sources involves multiple fields (e.g., fields of production and life), multiple levels (e.g., global, state, and city levels) and multiple angles (e.g., the views of macro-management, production technology improvement, and regional energy gradient utilization). Therefore, based on the structure of China's energy consumption, students should first review the key energy-using field or industry (e.g., industry, building industry, or iron and steel industry) and then choose a key energy-using area based on the distribution of our country's energy consumption (e.g., a specific major city with relatively high energy use, a specific area with typical significance both in energy consumption and pollutant discharge, or an area of significant concern such as the Beijing-Tianjin-Hebei region). Second, confirm the content to be researched, for example, a measure to improve energy utilization, e.g., a technical or management measure. The study can be implemented by analyzing energy utilization influence factors or analyzing performance characteristics of energy utilization and selecting several corresponding typical indexes as quantification measures. Finally, through horizontal regional comparison within a corresponding period or historical comparison of the same area, changing factors, the causes of good/bad differences in energy utilization and internal influence factors are analyzed to posit improvement measures. The screening process for topics is shown in Fig. 7.9.

Recommended topics for research are as follows:

(1) Analysis of the improved technical measures of energy utilization in a specific industry;
(2) Improvement suggestions for energy consumption in a specific city;

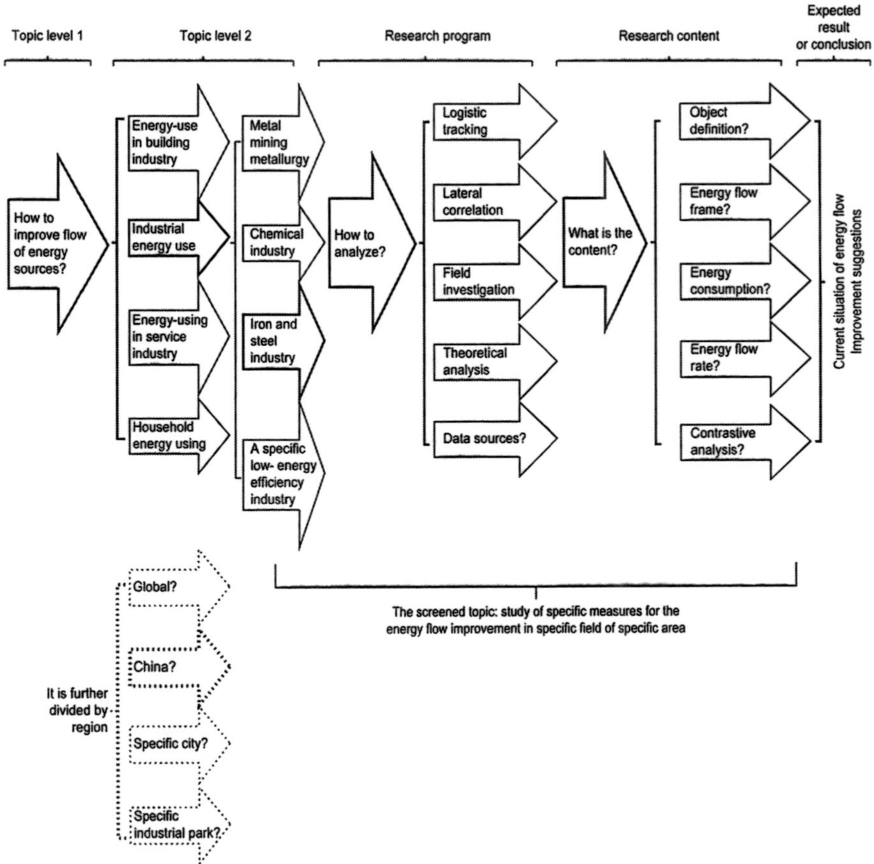

Fig. 7.9 Screening process of classroom topics discussion and analytical framework of discussion process

(3) Study of the optimization method of energy utilization in our country; and
(4) Study of the efficient utilization incomes of the energy in a specific industrial park.

Homework and Presentation in Class

Specific to the topics discussed in class, locate a related paper to share with the class via the group.

Recommended and Referenced Literatures

1. Dassisti, M., and L. Carnimeo. 2013. A small-world methodology of analysis of interchange energy-networks: The European behaviour in the economical crisis. *Energy Policy* 63: 887–899.
2. Baka, J., and R. Bailis. 2014. Wasteland energy-scapes: a comparative energy flow analysis of India's biofuel and biomass economies. *Ecological Economics* 108: 8–17.
3. Mao, J.S., Y.C. Du, and L.Y. Xu, et al. 2011. Quantification of energy related industrial eco-efficiency of China. *Frontiers of Environmental Science & Engineering in China* 5(4):585–596.
4. Mao, J.S., Y.C. Du, and H. Cai, et al. 2010. Energy efficiencies of industrial sectors for China' s major cities. *Procedia Environmental Sciences* (2):781–791.
5. Mao Jiansu, Zeng Run, and Du Yanchun, et al. 2010. Eco-efficiency of industrial sectors for China. *Environmental Science* 31(12):269–275.
6. Zhifeng, Yang, and Liu Jingling. 2010. *Introduction to Environmental Science*, 2nd ed. Beijing: Higher Education Press.
7. Bo, Li. 2003. *Ecology*. Beijing: Higher Education Press.
8. Enger, E.D., and B.F. Smith. 2004. Environmental science-cross subject. 9 ed. Beijing: Tsinghua University Press.

8 Value Flow of a Circular Economy

8.1 Presentation of Core Issues

The anthropogenic flow of material was thoroughly discussed in Chaps. 4 through 6; specifically, the basic process of material circular flow was analyzed to obtain the basic laws of ecological efficiency, circulation rate, yield change, and other parameters that change during the circular flow process. In this chapter, the core is still the circular flow of material, but the emphasis shifts to the exploration of the change in value in the anthropogenic flow of material and how value flows in a circular economy.

To obtain the value flow of a circular economy based on basic economic concepts, first, using the material as the carrier of value, attempt to develop the internal relation between material and value and build the value flow relationship under a circular economy. Second, develop a value flow relationship to explore the possible resource, environment, social and economic benefits of the implementation of a circular economy. Finally, propose a topic for further discussion.

8.2 Review of Several Basic Concepts of Economics: What Is Value

There are various of definition of "value" in economics, among which only the part that materials or energy shows service to human society in economic way is emphasized in the present chapter, it is called economic value. Although the definition of commodity (or service) varies between different schools in economics, "economic value" is used to measure the income a product (or service) brings to its producer or provider; generally, it is represented by an amount of cash.

In the value theory of economics, the value of a product (or service) can be embodied in the buyers and sellers in the market negotiating for the product (or service). For the buyer, it is the fee paid for the specific commodity, and for the

J. Mao et al., *Circular Economy and Sustainable Development Enterprises*,
https://doi.org/10.1007/978-981-10-8524-6_8

seller it is the cost paid for the delivery of the specific commodity. There are many related terms in application; however, in this chapter, we only provide a brief summary of the terms relevant to the contents and concept of this chapter.

Value in use refers to the property capable of meeting a specific human need, e.g., a house protects from wind and rain and guarantees security. In the analysis of the anthropogenic flow of material, this type of property is represented by its usability by human beings.

Market value refers to the social value of a commodity based on the labor required for its production. Market value refers to the price of an asset in a trading market and is the price a voluntary buyer pays to a voluntary seller under the condition of rational conduct without any compulsion.

Market price refers to the transaction price of a commodity in a market. Generally, there is only one price for the same commodity in the same market. The market price is formed via the internal competition of a department and depends on the value of the commodity and the supply–demand relationship of the commodity in the market. Price is the monetary expression of the commodity in the market transaction process. In the value flow analysis of this chapter, the value is measured by the price of the product.

Added economic value is called economic value added. It is equal to the sum of after-tax and before-interest profit minus all capital costs. It is the surplus revenue after deducting all costs and, to some extent, reflects the earned profit of the enterprise. This concept can be used to analyze the economic benefit of the material circular flow to the enterprise.

The circular flow of value described in this chapter refers to the flow of economic value that occurs with the anthropogenic circular flow of material in a composite economic system of human society. Within the circular flow of value, the value occurs during the period when the material waste produced in production and use are treated as secondary resources for the upstream life cycle. In this manner, it differs from the service value of material in a natural ecosystem. The relevant keywords are serviceability of material, product or service, transaction, price or value, and profit of enterprise.

8.3 Method to Build the Value Flow Framework in a Circular Economy

8.3.1 Method to Build the Relationship Between Material and Value

(I) Basic idea

As previously mentioned, the anthropogenic flow of material is to meet the specific needs of human beings, and it is the final product developed by means of a specific service function. From natural resources to final products, the material must undergo many transformations, such as shape, function (service performance), and regional space. These transformations not only require the power of energy purchased from other departments, but their ability to serve human beings also improves constantly as the service value of the material transforms through a series of major changes. The change in material service value is embodied in the economic system to form the flow of value.

Material is the carrier of value; thus, with the circular flow of material, there is a concurrent circular flow of value. Furthermore, there is a specific necessary connection between the circular flow of value and the circular flow of material. Based on this, if the relationship between material and the value it carries can be established, we can obtain the flow relationship of the value by means of the circular flow of material.

To establish the quantitative relationship between material and its value, several points should be considered. ① As seen from the contents of Chaps. 4 through 6, the circular flow of material is inseparable from the circular flow of material in a product life cycle, and in different life cycle periods, the product has different values. ② A product consists of materials. If the value of the product is shared with its components, each component in the product becomes the carrier of value, and in different life cycle periods of the product, the material has different values. ③ A product always consists of many materials. If the product is represented by a specific material M, the circular flow process in the product life cycle is simplified to the circular flow process of element M, the value of the product is only apportioned to this material element, and the value of the product in each life cycle evolves into the value of material element M in its corresponding life cycle. Thus, the certain inevitable relationship between material and value might be established.

(II) Price of element M

The price of material element M in this type of life cycle is defined as the ratio between the value of a product in a certain life cycle and the mass of material element M:

$$V = \frac{E}{\mathrm{M}} \tag{8.1}$$

where V is the price of material M in Yuan/t, showing the quantity of value carried by the unit mass of material; E is total the value of a specific material flow in Yuan; and M is the quantity of material M in the material flow in t. For example, an iron and steel plant produces 10,000 t of steel products, valued at 15 million Yuan, and based on this, it can be calculated that the average price of Fe in the steel production stage is 1500 Yuan/t.

In the production process of a product, the price of element M reflects the processing degree of the industrial system to element M. The higher the price of element M, the higher its processing degree and the higher its industrial added value. There is a more obvious example, specifically, the price of element M in metal products (e.g., steel, aluminum and copper) is generally higher than the price of element M in corresponding metal mining powder, but lower than the price of element M in metal products (e.g., steel cylinder, aluminum pot, and copper pipe).

In addition, the price of element M is also influenced by the product type, life cycle phase, region, technical equipment, and management level. For example, the price of Fe in auto parts may reach several thousand or even ten thousand Yuan, while the price of Fe in a cast iron sewer line is usually approximately 1000 Yuan.

To clarify the relationship between the human economy and the resource environment and economy due to material flow, the following assumptions are made. ① If the material is beneficial to the service object, its service value is positive; otherwise, if it is harmful or unfavorable to a specific service object, its service value is negative. ② Human beings and the natural element have the same existence value; therefore, the material serves human beings and nature identically. Under this assumption, when human beings obtain element M from natural resources, material M shows serviceability to human beings and the value of its anthropogenic flow is positive. If the natural resource environment and environmental system are damaged, a resource-use fee must be paid to the system to compensate for the loss of natural resources or measures must be taken to repair the damaged natural system. When human beings discharge wastes into the environment, the material has lost its usability and its anthropogenic flow value is negative. If the external environment is also damaged by a portion of the flow, human beings should pay a sewage charge to the environmental system to compensate for the loss caused by the decrease in environmental quality, or should take corresponding measures to assist in the recovery of the environmental system by means of an environmental governance department. In this situation, the price of element M in the discharged material flow should be negative. Otherwise, if the discharge of element M into the environment is beneficial to the external environment, the industry department could receive capital from the department that benefits by selling waste resources or could save capital via exemption of a sewage charge. In this situation, the price of element M in a waste material flow should be positive.

(III) Change in the price of element M during the life cycle

In the product life cycle, with the implementation of industrial production, raw materials will undergo a series of physical–chemical changes and be transformed into different industrial products and the price of element M increases by varying degrees. During product use, the composition of the product wears constantly, the product's service ability gradually reduces, and the price of element M gradually declines. The change in the price of element M refers to the phenomenon in which the price of element M constantly changes with the life cycle phase of the product. Based on this, the change in the price of element M with the product life cycle phase of the product is called the change map of the price of element M in the life cycle, as shown in Fig. 8.1. It is simple to observe that the change in the price of element M only occurs with the change of phase of the life cycle.

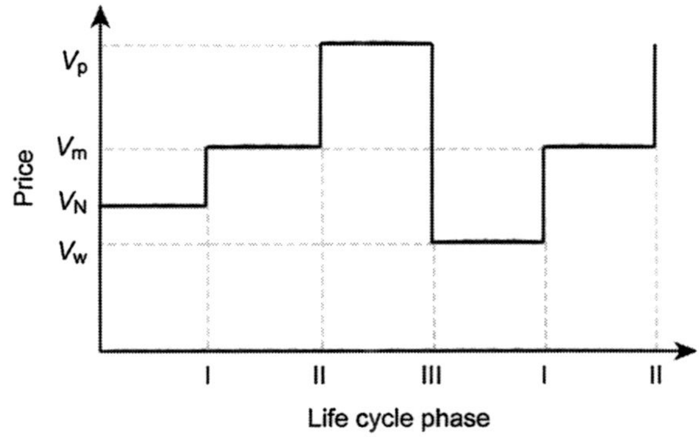

Fig. 8.1 Change of the element M price in a life cycle reprinted from Mao and Lu [5], with permission from Journal of Materials and Metallurgy (Chinese)

Under normal circumstances in a product life cycle, element M exists in the four forms of natural resource, industrial material, product and discharged waste; accordingly, the price of element M is divided into the following four types:

(1) The price of element M in natural resources is represented by V_N;
(2) The price of element M in industrial materials is represented by V_m;
(3) The price of element M in a product is represented by V_p; and
(4) The discharged wastes are regarded as the waste resources, and the price of their element M is represented by V_w.

In the case of lead-acid batteries, lead refining, manufacturing of lead-acid batteries, use and other processes constitute the different stages of its life cycle. According to statistical data from China's Foreign Trade Department in 1999, we can calculate the price of lead in different stages in the units of USD/t. Thus, we can see the change in Pb price in different life cycle phases.

8.3.2 Method to Build a Value Flow Framework

As previously mentioned, material is the carrier of value and, after building the relationship between material and value, we can establish the value flow framework under a circular economy by means of a circular flow of material.

(I) Circular flow of element M

The life cycle of an industrial product generally includes its manufacture, use, and scrap waste reclamation. It is a very long material flow process. If element M is a key component of a product, we can develop the flow diagram of element M in a product life cycle, as shown in Fig. 8.2. The I, II, and III in the figure represent the three stages of the product life cycle.

Please note that the quantity of the above materials reflects the content of element M.

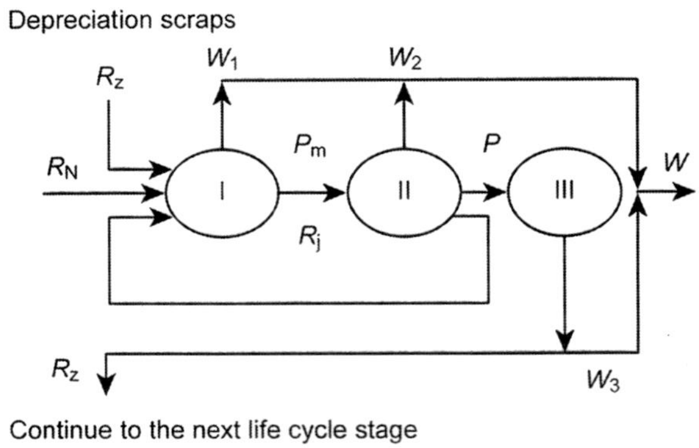

Fig. 8.2 Flow diagram of element M in product life cycle stage I manufacture of industrial material, stage II product manufacture, and stage III product use. *Note* R_N—input of natural resources, t; R_z, R_j —input of depreciation wastes and processing wastes, t; P_m —input of the raw materials in the stage of product manufacture, and at the same time, it is the output capacity of the industrial materials in production stage, *t*; *P*—product output, t; W_1, W_2, W_3—quantity of the discharged wastes and pollutants in the production stage of industrial materials, product manufacture stage and after scrap of product use, *t*; *W*—total discharge amount of element M, $W = W1 + W2 + W3$, *t*. Reprinted from Mao and Lu [5], with permission from Journal of Materials and Metallurgy (Chinese)

From Fig. 8.2, it is simple to observe that in the flow process of element M, there is not only a one-way flow from natural resources to the environment, but there is circular flow.

(1) General circulation—After the scrap of industrial products, a portion of element M returns to the production stage of industrial materials as raw material for recycling. In Fig. 8.2, I represents the flow of depreciation waste R_z.
(2) Medium circulation—In the wastes produced during the product manufacturing stage, a portion of element M returns to the production stage of industrial materials as raw material for recycling. In stage I of Fig. 8.2, the flow of processing of waste R_j is shown.

In addition, there is a small circulation in the wastes produced during a life cycle phase. A portion of element M returns to the same period of the life cycle for recycling, showing that the wastes from downstream procedures in the enterprise return to upstream procedures as raw material for recycling. As a small circulation is a cycle of the material conducted during the same period of the life cycle, it is not labeled in Fig. 8.2.

In terms of the entire product system, due to the existence of the three circular flows, the quantity of the industrial products output from the unit of natural resource significantly increases, and the quantity of wastes and pollutants discharged to the environment obviously decreases. Put differently, resource efficiency and environmental efficiency improve greatly; therefore, the existence of the three circular flows has an effect on saving natural resources and protecting the environment.

(II) Circular flow of value

It is not difficult to understand that in a product life cycle, with the circular flow of each element M, there is a circular flow of value; furthermore, the flow of each value flow should be equal to the value of the mass flow rate of element M multiplied by the price of element M in the corresponding material flow life cycle. Based on the flow of element M in Fig. 8.2 and the price of element M in Fig. 8.1, we can draw the value flowchart of element M in the product life cycle, as shown in Fig. 8.3. The figure can be considered the circular flowchart of value in a product life cycle.

It can be observed from Fig. 8.3 that in stage I, with the input of raw materials, there are three types of value inflows, namely $R_N V_N$, $R_z V_w$, and $R_j V_w$, in which $R_z V_w$ and $R_j V_w$ occur with the circular flow of element M, and its existence will substitute the inflow of a certain amount of value related to natural resources. For example, when natural resources and waste resources have the same yield of element M, the circular flow of these two values will substitute the value flow with $(R_z + R_j) \cdot V_N$ related to the natural resources, thus saving the value in the quantity of $(R_z + R_j) \cdot (V_N - V_w)$. In stage II, the value flow $R_j V_w$ occurs with the medium circulation of element M, and its existence not only reduces sewage charge $R_j V_w$ but also increases the output of the value with the quantity of $R_j V_w$; thus, there is an income increase of $2R_j V_w$. In stage III, the value flow $R_z V_w$ (which is not always same as the value inflow in stage I) occurs with the major circulation of element M and its existence will reduce the consumer's sewage charge $R_z V_w$. The sale of waste resources allows consumers to recover $R_z V_w$ capital, and the two items are summed to improve economic income by $2R_z V_w$. In summary, for the product producer and customer (user), the circular flow of value decreases expenditures and increases income.

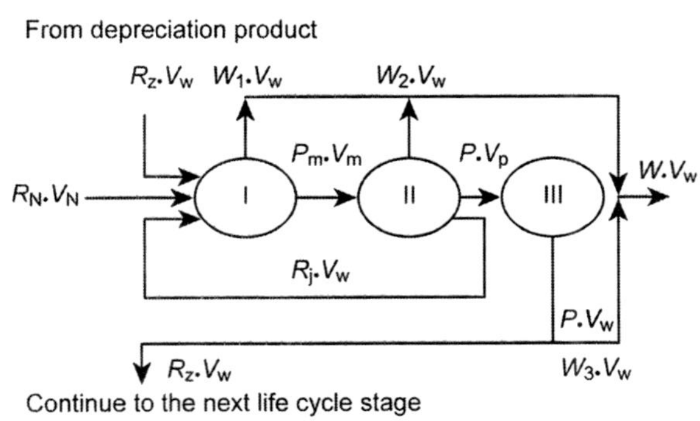

Fig. 8.3 Circular flowchart of value in product life cycle Reprinted from Mao and Lu [5], with permission from Journal of Materials and Metallurgy (Chinese)

(III) Case analysis of a value circular flow

The 1999 lead-acid battery system continues to be used as an example, in which Pb is selected as the representative substance of a lead-acid battery.

Considering the years of study, there is no fair pricing for mineral resources, wastes, or pollutants discharged into the environment; therefore, there is no method to calculate the price of element M in natural resources and wastes. To verify the effectiveness of this theory in this chapter, in the example below, the starting point of the lead-acid battery system returns to the lead concentrate obtained from ore dressing. The lead concentrate is used as a substitute for the lead ore resource, and reclaimed scrap lead is used as a substitute for the scrap lead generated in the lead-acid battery product system; it is considered to be the price of Pb in lead ore resources and wastes. As the Pb yield from an ore dressing period in 1999 was approximately 83.8%, the lead concentrate placed into production was approximately 281.77 kt in 1999. Accordingly, the environmental discharge from lead smelting production is exclusive of the amount of lead lost during ore dressing; thus, it changes to 34.43 kt from 88.90 kt in Fig. 5.2. Therefore, the total environmental discharge from a lead-acid battery system decreases to 250,700 t from the original 305,200 t. In this case, the flow diagram of the Pb in a lead-acid battery life cycle in 1999 evolves into Fig. 8.4. In the figure, the units remain kt.

Then, based on the price of Pb in various product life cycle phases, such as lead refining, lead-acid battery manufacturing and use, $V_N = 150.62$ USD/t, $V_m = 468.19$ USD/t, $V_p = 2\ 381.45$ USD/t, and $V_w = 145.94$ USD/t. Based on the Pb flow diagram in Fig. 8.4 of the lead-acid battery, the lead flow within each material flow is multiplied by its price in the corresponding life cycle phase to obtain the circular flowchart of the value of a lead-acid battery life cycle, as shown in Fig. 8.5. In the figure, the units are millions of USD.

Compared with scrap lead, the price of Pb in lead concentrate is higher (4.68 USD); thus, it can be calculated that due to the circular flow of Pb in stage I, 103,400 t (specifically, 9.09 + 1.25) of lead concentrate will be saved and 483,900 USD (103,400 t × 4.68 USD/t) of expenditure will be saved. In stage II, due to the existence of the circular value flow of 1.824 million USD, there is 3.648 million

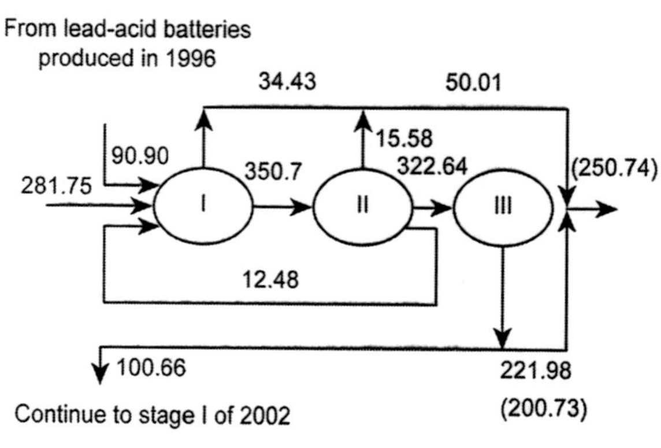

Fig. 8.4 Flow diagram of the Pb in lead-acid cell life cycle in China 1999, *Note* the starting point is lead concentrate. Data unit is kt. *Source* Mao [16]

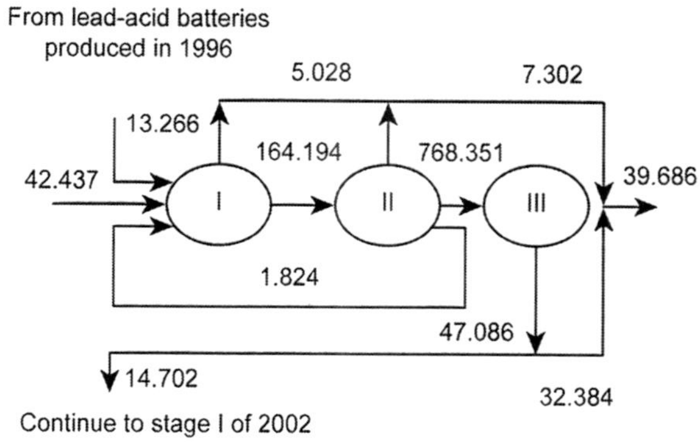

Fig. 8.5 Value flow of lead-acid cell life cycle in China 1999 (unit is million USD) *Source* Mao [16]

USD of increased income (if it is discharged into environment, there is a 1.824 million USD sewage charge). In stage III, if 31.2% of waste batteries can be recycled, a 14.70 million USD circular flow would form; thus, it is available to assist consumers in recovering 29.40 million USD of capital (if it is discharged into the environment, there is a 14.70 million USD sewage charge). Therefore, it can be seen that a material's circular flow can bring considerable economic benefits to the producer and consumer.

The above value circular flow has the following features. ① It is established on the basis of a material's circular flow combined with the material's circular flow laws, especially for the dynamic loop law of material, and is available to study the dynamic activity pattern of an economy. ② The environment beyond the economic system is treated as an integral whole; therefore, regardless of whether element M occurs in natural resources or in wastes, it has a specific value that embodies the economic relationship between the industrial system and the natural environment. ③ The difference in element M's price in an industrial system and as a waste resource indicates the difference between these two resources' use of element M, and this difference is the basis of the analysis of the relationship between waste resource utilization and economic benefit. ④ As element M in resources (including natural and waste resources) has a certain value, on the basis of the value circular flow, the economic means are applied to control the resources and the environment.

Currently, there is a lack of fair pricing of natural resources and environmental pollutants in economics; therefore, in the analysis of the value flow, it is difficult to extend the condition where a natural resource is regarded as a starting point and the environment is regarded as an end point. Furthermore, in the practical use of different types of wastes, the price of an element varies based on the different service abilities of the material. For example, the lead in battery scrap is easy to recycle and has better utility value; thus, it is more expensive, while the lead fumes or lead dust dissipated into the atmospheric environment after heating during the production process, although they may contain lead, not only cannot be recycled but also seriously threaten human health; therefore, their price should be assigned a higher negative. There are several details to be quantitatively refined with the further study. The circular flow of value within a material's circular flow will bring far-reaching influence to the resource, environmental, social, and other branches of economics.

8.4 What Are the Earnings of a Circular Economy?

8.4.1 How Are the Earnings Defined?

(I) Basic idea

The flow of material in a natural ecosystem is standardized by a circular economy by imitating the closed circular flow of materials in the human economic system; its purpose is to reduce the influences of human beings on the environment and to achieve harmonious development between human beings and the natural ecosystem. Currently, there is an inadequate study of relevant problems; for example, how does material flow produce income, what type of income does it produce, and how much income does the material flow produce? No detailed report currently exists. These problems are the key elements to determine whether a circular economy can be smoothly implemented, especially during the current stage of a developing country. Therefore, a theoretical discussion of the influences of a material circular flow on the environment, analysis of the various incomes produced, and a verification of its scientificity will not only solve the above problems, but will promote the further development of a circular economy.

The core of the circular flow is material circular flow, which has substantial differences from the linear flow of material in a traditional economy; linear flow refers to the "natural resources-product-environmental emission" flow. If a comparison is made between the material circular flow in a circular economy and the material linear flow in a traditional economy, while considering that material is the carrier of value, within the circular flow of material there is a circular flow of value. By means of the value flow framework under a circular economy established in Sect. 3 of this chapter, we identify the differences between a material's circular and linear flows and identify the difference in natural resource consumption caused by these differences, environmental pollution discharge, social services, and economic benefit. Through the above identification process, we should be able to obtain the resource income, environmental revenue, social income, and economic benefit brought thereby.

(II) Definition of earnings

It is not difficult to observe from Fig. 8.2 that the relationship between the entire product system and the external environment is presented as follows: ① to provide products to society to meet a specific service need and provide social services, represented by the product's annual output P with unit t/a; ② to remove materials from natural resources to form the resource load of element M, represented by R with unit t/a; and ③ to discharge wastes and pollutants containing element M to form a discharge load, represented by W with unit t/a.

Compared with the material's linear flow, the material's circular flow can not only bring changes from a material perspective, but also it brings changes from the perspective of social services and the economy. For this purpose, the resource income, environmental revenue, service income, and economic benefit of the material's circular flow can be defined.

1. Earnings from a material perspective

For the different forms of material flow, to obtain same quantity of product (or service), the resource load is different under a material's circular and linear flows; the difference is the amount of natural resources saved by the material's circular flow, called the resource income and represented by ΔR. For the same reason, to obtain the same quantity of product, the discharge load of these two material flows is different; the difference is the decrement of waste emission brought by the material's circular flow, called environmental revenue and represented by ΔW.

From another perspective with the same resource or discharge load, the product output is also different under a material's circular and linear flows; the difference is the increment of product output caused by the material's circular flow, called revenue and represented by ΔP. The increment of product output corresponding to the same amount of resource load is called the service revenue related to resources, and the increment of product output corresponding to the same amount of discharge load is called service revenue related to the environment.

It is not difficult to observe that several earnings above are revenues from a material perspective caused by a material's circular flow, and they belong to an absolute index.

In application, to reflect the essence of the earnings caused by a material's circular flow, the influences of a production scale must be removed. For this purpose, it is defined that the relative index of corresponding earnings is represented by the ratio between the absolute index of the earnings above and the product output during the same period (represented by material content, P).

2. Earnings from an economic perspective

From an economic perspective, the production department provides the "value increment" and can be regarded as the source of value.

To analyze the economic benefits of the value circular flow brought to the production department, the difference between the value outflow and inflow in a specific production phase is defined as the value increment in the period, represented by ΔV_i with unit t. Subscript i represents the product life cycle phase. It is not difficult to understand that net earnings of the production department should be equal to the value of the value increment in the corresponding life cycle phase minus the corresponding values of various manpower, equipment, and other inputs. The larger the value increment is, the larger the corresponding net earnings.

By comparing Figs. 8.4 and 8.5, it is simple to observe that due to the differences in the material flow status and the positive and negative values of the waste flow, there are differences in the value increment in the production phases. The difference between the value increment under a material circular flow and one under a material linear flow is defined as the economic benefit brought to the corresponding production department via the material's circular flow, represented by $\Delta(\Delta V)_i$ with units Yuan/a. It is not difficult to observe that this indicator belongs to an absolute index. If the influence of production scale on economic benefit is not considered, the ratio between the absolute index of economic benefit and product output, namely $\Delta(\Delta V)_i/P_\tau$, is regarded as the index to use to measure economic benefit.

Additional information follows: If all wastes are effectively reused under a material's circular flow, one portion is used internally by the product system and another portion is utilized by other departments beyond the product system, and the value of the waste material flow is positive. In the case of a material's linear flow, all wastes are discharged to the natural environment, and the value of the waste material flow is negative. In addition, the case in which the price of element M varies within the different types of wastes is not considered in this study.

8.4.2 What Amount of Earnings Does a Material's Circular Flow Provide?

To make clear the amount of income provided by a material's circular flow from a material perspective and to associate the basic laws of the material circular flow obtained previously, we follow the analysis framework of the material's anthropogenic flow via the tracking method established in Chap. 5 and compare the differences from the perspective of material flow quantity under the conditions of circular and linear flows.

(I) Flow diagram of element M in a product life cycle

It is still assumed that a specific material element M in a product is used to represent the product. In this case, the material flow process in a product life cycle evolves into the anthropogenic flow process of material element M, so as to establish the organic relations between the basic laws of material flow in this chapter and Chap. 5. Based on this concept and specific to element M in the product and assuming the material inflow in each life cycle phase is equal to outflow, we can draw a material flow diagram to reflect the material flow amount in each product life cycle phase (called the flow diagram of element M in product life cycle phases), as shown in Fig. 8.6.

The following is assumed in Fig. 8.6: ① the average service life of a product is $\Delta\tau$ years, and it is constant within the time frame of the study; ② the duration of each production process is not considered; and ③ the product is completely scrapped after production year $\Delta\tau$ to form depreciation scrap; furthermore, the product returns to material production for recycling the year it is scrapped. It is not difficult to imagine that the figure is completely consistent with Fig. 5.1. The difference is that Fig. 8.6 generally refers to various products and element M is the representative element in the product and that Fig. 5.1 is specific to lead-acid batteries and Pb is used as the example. In Fig. 8.6, each symbol is also consistent with Fig. 5.1, referring to the quantity of the representative element M in the product. It is displayed again here for the convenience of contrastive analysis under the condition of material linear flow.

(II) Material linear flow

It is not difficult to imagine that if all wastes produced in the product system are non-recyclable, they would be directly discharged into the environment. In this case, the industrial flow of material is presented as a linear flow flowing through the product system from natural resources to the environment, and the flow diagram of element M in the product life cycle is also simplified, as shown in Fig. 8.7.

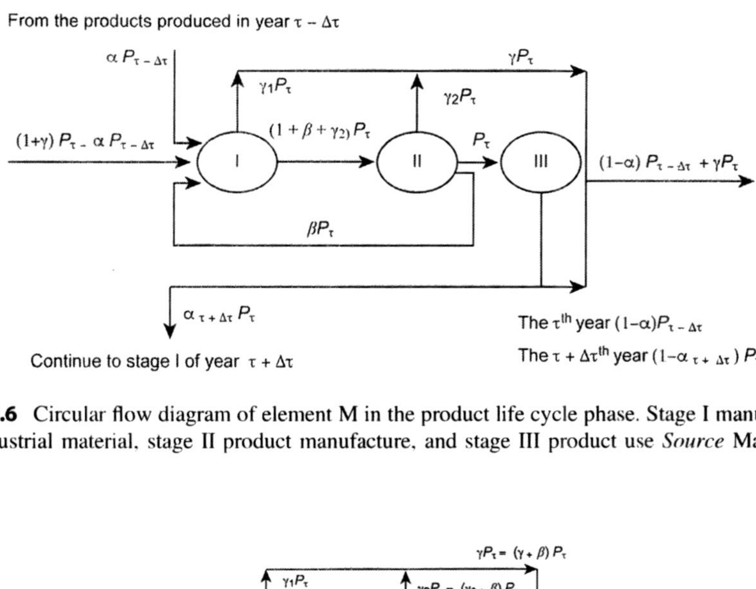

Fig. 8.6 Circular flow diagram of element M in the product life cycle phase. Stage I manufacture of industrial material, stage II product manufacture, and stage III product use *Source* Mao et al. [12]

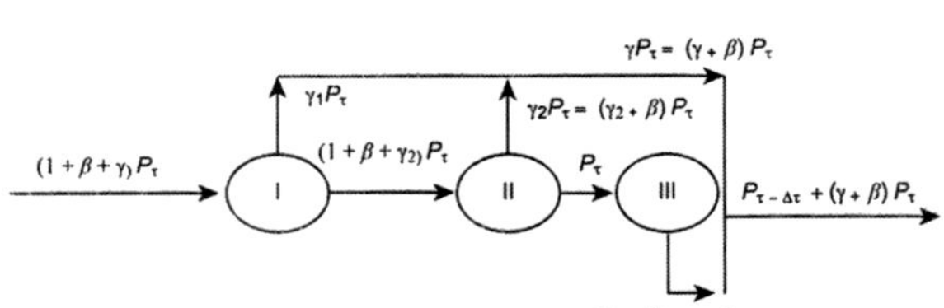

Fig. 8.7 Linear flow diagram of element M in product life cycle *Source* Mao et al. [12]

For the convenience of the earnings analysis in the following subsection, Fig. 8.7 is drawn with the same amount of product output (namely P_τ) as in Fig. 8.6, and it is assumed that the utilization percent of different materials (including natural and waste resources) in each production process is consistent with the corresponding values in Fig. 8.6.

(III) Earnings from a material perspective

Within the different material flows, to obtain the same quantity of product, the resource load is different under the material's circular and linear flows. The difference between the two is the amount of natural resources saved by the material's circular flow, called resource income and represented by ΔR (unit is t/a). For the same reason, to obtain the same quantity of product, the discharge load of these two material flows is different, and the difference is the decrement of waste emission from the material's circular flow, called environmental revenue and represented by ΔQ (unit is t/a).

From another perspective, with the same resource or discharge load, the product output under the material's circular and linear flows is also different, and the difference is the increment of product output from the material's circular flow, called the service revenue, represented by ΔP (unit is t/a). The increment of product output corresponding to the same amount of resource load is called the resource-related service revenue and is represented by $\Delta P_{\tau R}$. The increment of product output corresponding to the same amount of discharge load is called the environment-related service revenue and is represented by $\Delta P_{\tau Q}$.

It is not difficult to observe that the earnings above are the revenues from a material perspective from the material's circular flow and belong to an absolute index.

In application, to reflect the essence of the earnings caused by a material's circular flow, the influences of the production scale must be removed. For this purpose, it is defined in this chapter that the relative index (shown in Table 8.1) of corresponding earnings is represented by the ratio between the absolute index of the earnings above and the product output (P_τ).

Table 8.1 Earnings from a material perspective from the circular flow of material

Category of earnings	Resource income ΔR	Environmental revenue ΔQ	Service revenue	
			Resource-related	Environment-related
Absolute index	$\beta P_\tau + \alpha P_{\tau-\Delta\tau}$	$\beta P_\tau + \alpha P_{\tau-\Delta\tau}$	$(r - r')R$	$(q - q') \cdot Q$
Relative index	$\beta + \alpha p$	$\beta + \alpha p$	$1 - \frac{r'}{r}$	$1 - \frac{q'}{q}$

Note In the table, $r = \frac{1}{1+\gamma-\alpha p}$, $r' = \frac{1}{1+\beta+\gamma}$, $q = \frac{1}{\gamma+(1-\alpha)p}$, $q' = \frac{1}{p+\beta+\gamma}$, $p = \frac{P_{\tau-\Delta\tau}}{P_\tau}$, and the relative index are the ratio between the absolute index and P_τ

(IV) Discussion of material earnings

1. Resource income and environmental revenue

It can be observed from Table 8.1 that the resource income from the material circular flow or the environmental revenue is the sum of the quantity ($\alpha P_{\tau-\Delta\tau}$) of the material's major circulation and the amount (βP_{τ}) of medium circulation in the product system. The larger the quantity of material being recycled, the higher is the resource income or environmental revenue.

From the perspective of the relative index of these two earnings (the two earnings are related to a large circulation rate and the output variation ratio in the product life cycle phase), the higher the circulation rate is, the higher the resources income and environmental revenue; furthermore, the decline in production is beneficial to obtain improved resources and environmental revenue.

2. Service revenue

It can be observed from Table 8.1 that the resource-related service revenue from the material's circular flow is the function between the difference value ($r - r'$) of the resource efficiency and resource load (R) under the material's circular and linear flows. Here, resource efficiency refers to the resource load per unit of product output from the product system. It is represented by r (unit is t/t), is related to the flow status of material, and is the function of the major circulation rate, emission ratio, and product life cycle yield ratio (as shown in the notes of Table 8.1). Furthermore, the improvement of the major circulation rate, decrease in the emission ratio, and decline in production are conducive to the improvement of resource efficiency. From the perspective of a relative index of resource income, it is only related to the resource efficiency under the flow of these two materials. Under a material circular flow, the higher the resource efficiency is, the larger the relative value of the resource-related service revenue. Therefore, the obtained resource-related service revenue is higher.

There is a similar conclusion for the environment-related service revenue from a material's circular flow. The difference is that under this circumstance, the original resource efficiency and resource load are replaced by environmental efficiency (represented by q) and discharge load, respectively; furthermore, an increase in production is beneficial for the improvement of environmental efficiency.

8.4.3 How Much Economic Benefit Does a Value Circular Flow Bring?

A similar concept is used to analyze the economic benefit brought by a material's circular flow by comparing the difference between material circular and linear flows from an economic perspective.

(I) Circular flow diagram of value in a product life cycle

It is not difficult to understand that in a product life cycle, with the flow of each material, there is a flow of value; furthermore, each value flow should be equal to the value of the mass flow rate of element M multiplied by the price of a unit of element M in the corresponding material flow of the life cycle. On the basis of the material circular flow (Fig. 8.6) and linear flow (Fig. 8.7) and combined with the price change of element M during the life cycle (Fig. 8.1), we can obtain the flow diagram of value in a product life cycle, as shown in Fig. 8.8.

(II) Value flow under a material's linear flow

Using the same concept employed in Sect. 3 of Chap. 5, the quantity of each material under a linear flow is multiplied by the element price in the corresponding stage (Fig. 8.1) to obtain the flow diagram of the value flow in a product life cycle under a material linear flow, as shown in Fig. 8.9.

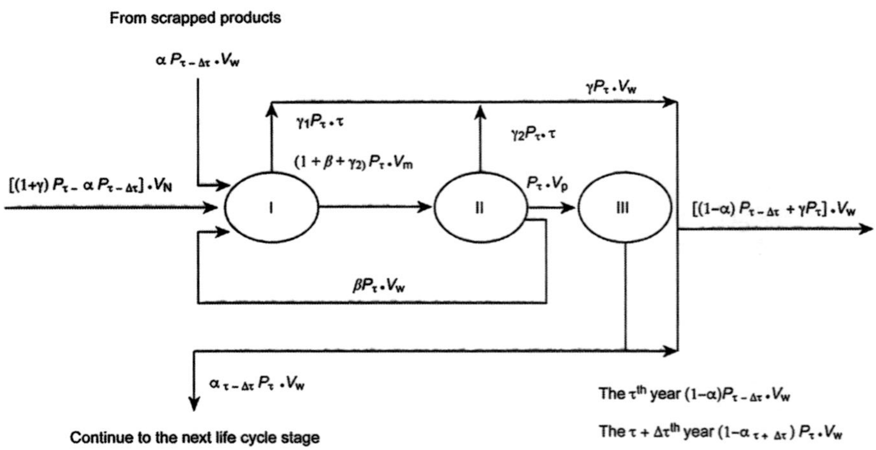

Fig. 8.8 Circular flowchart of value in a product life cycle *Source* Mao et al. [12]

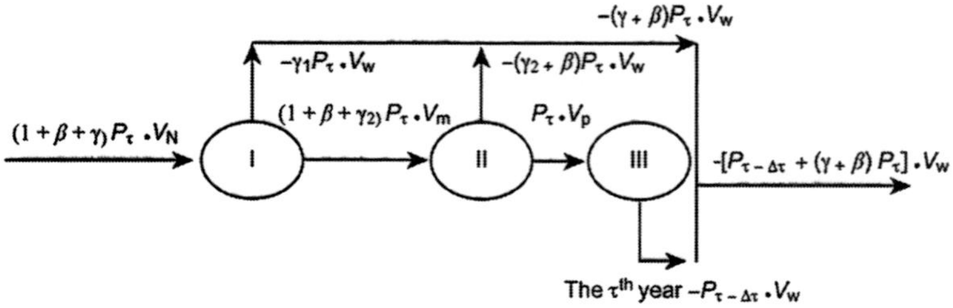

Fig. 8.9 Value flow diagram under material linear flow *Source* Mao et al. [12]

In this case, the wastes produced by the product system are totally discharged into the natural environment, and it is assumed in this chapter that these wastes would have an adverse impact on the environment; therefore, an environment compensation fee must be paid to the environmental protection administration and the value of the waste flow would be negative.

(III) Earnings from an economic perspective

By comparing the value flow (Fig. 8.8) under material circular flow and the value flow (Fig. 8.9) under material linear flow, specific to the production phase of industrial materials and the processing stage of products, the change in economic value under material circular flow is analyzed to obtain the economic benefit in the corresponding life cycle phase, as shown in Table 8.2.

(IV) Discussion of economic benefit

In the production phase of industrial raw materials, the economic benefit brought by the material's circular flow is related to the element of material circular flow and the price difference ($V_N - V_w$) of element M between natural resources and wastes. If the price of natural resources is higher than the price of waste resources, the resource efficiency is higher. The higher the medium circulation rate or the lower the waste emission is in a stage, the higher the economic benefit in that stage. This is because higher resource efficiency, medium circulation rate, or smaller waste emission ratio indicate better material utilization in the production phase of industrial raw materials; concurrently, additional low-price waste resources are substituted for high-price natural resources, bringing increased economic benefit. Furthermore, the larger the price difference ($V_N - V_w$) is between natural resources and waste, the higher the obtained economic benefit. If there is a difference in the price of element M between natural resources and waste (i.e., $V_N < V_w$), with the improvement of resource efficiency or medium circulation rate, or with the drop of the emission ratio of wastes in a life cycle stage, the net earnings would constantly reduce. Under these circumstances, although the utilization of wastes can still protect natural resources and reduce waste emission, there is no economic benefit. This type of situation primarily occurs in ① an area with a shortage of waste resource, e.g., the majority of the developing countries and ② an area without enough reasonable resource policies, e.g., a country where the subsidy policy of scarce resources is still being implemented. This situation must be corrected by improving resource policies.

Table 8.2 Economic benefit of material circular flow

Category of earnings	Production phase of materials	Product manufacturing
Absolute index	$\left(1+\beta-\gamma_1-\frac{1}{r}\right)\cdot P_\tau\cdot(V_N-V_w)$	$(\beta+3\gamma_2)\cdot P_\tau\cdot V_w$
Relative index	$\left(1+\beta-\gamma_1-\frac{1}{r}\right)\cdot(V_N-V_w)$	$(\beta+3\gamma_2)\cdot V_w$

Note In the table, $r=\frac{1}{1+\gamma-\alpha p}$ and the relative index are the ratio between the absolute index and P_τ

During product manufacturing, the economic benefit brought by a material's circular flow is primarily related to the price of element M in its waste resources, medium circulation rate, and waste emission ratio. In waste resources, the higher the price is of element M, the larger the economic benefit, and the higher the medium circulation rate and waste emission ratio is, the higher the economic benefit. This is primarily caused by assuming that the wastes discharged outside of the system are effectively utilized by another industrial sector.

8.4.4 Example Analysis—Analysis of the Earnings of a Pb Circular Flow in a Lead-Acid Battery System

(I) Research method

Pb is the main substance in a lead-acid battery, occupying 80% of the total mass; furthermore, Pb is from non-renewable resource lead ore, and it is an eco-toxic substance; it is critical in and the representative element of a lead-acid battery. Accordingly, in this study, China's 1999 lead-acid battery system is used as an example. It is combined with the circular flow status of Pb in a specific world-class Swedish lead-acid battery system between 1992 and 1995, and China's corresponding index is replaced by the index of Pb circular flow in Sweden to predict the earnings of Pb circular flow in China.

In the material flow, we use the data from Fig. 8.4, namely the indicators of the total Pb flow in China's lead-acid battery system in 1999: $P = 322{,}600$ t, $R = 281{,}800$ t, and $Q = 250{,}700$ t; furthermore, based on this flow, we calculate the resource efficiency and environmental efficiency under the service level based on product volume. The data are provided in Table 8.3. Using a similar method, the indicator of Pb flow in Sweden's lead-acid battery system is also calculated and presented in Table 8.3.

In the value flow, we use the price of Pb in a lead-acid battery life cycle obtained from the case in Sect. 3 of this chapter: $V_p = 2\ 381.45$ USD/t, $V_N = 150.62$ USD/t, $V_m = 468.19$ USD/t and $V_w = 145.94$ USD/t.

(II) Result and discussion

The data from the Pb circular flow elements and the Pb price obtained above are substituted into each equation in Tables 8.1 and 8.2 to calculate the various earnings under a circular flow. The results are presented in Table 8.4. In the forecasting section, the total Pb flow in the lead-acid battery system in China in 1999 is regarded as the calculation basis.

As observed in Table 8.4, the Pb circular flow saves Pb resources, reduces environmental emission of Pb, increases output of lead-acid batteries, and improves the economic income of the Pb and lead-acid battery production departments. For

Table 8.3 Comparison of Pb circular flow elements between China's and Sweden's lead-acid battery systems

	Resource efficiency r (t/t)	Environmental efficiency q (t/t)	Major circulation rate α (t/t)	Medium circulation rate β (t/t)	Emission ratio γ_1 (t/t)	Emission ratio γ_2 (t/t)	Total emission ratio γ (t/t)	Yield ratio p (t/t)
China	1.145	1.287	0.312	0.039	0.107	0.048	0.155	0.904
Sweden	85.47	85.47	0.990	0.123 6	0.002 655	0.000	0.002 655	1

Source Mao et al. [12]

Table 8.4 Earnings from examples of Pb circular flows of lead-acid battery systems

	Category of earnings	Resource income ΔR	Environmental revenue ΔQ	Service revenue		Economic benefit	
				Resource-related	Environment-related	Production phase of materials	Product manufacturing
Relative index	Current situation	0.321	0.321	0.268	0.292	0.274	26.707
	Prospect	1.114	1.114	0.990	0.990	5.191	18.038
Absolute index	Unit	kt/a	kt/a	kt/a	kt/a	Thousand dollars/a	Thousand dollars/a
	Current situation	103.57	103.57	86.47	94.21	88.40	8 616.75
	Prospect	359.29	359.29	319.41	319.41	1 674.82	5 819.78

Notes ① The relative index of the current earnings of the material circular flow is estimated by the Pb flow in China in 1999, namely $r = 1.145$, $r' = 0.838$, $q = 1.287$, $q' = 0.911$, and $p = 0.904$
② The relative index of prospective earnings is estimated by the Pb flow in Sweden between 1992 and 1995, namely $r = q = 85.47$, $r' = q' = 0.888$, and $p = 1$
③ The output of lead-acid batteries in China in 1999, namely $P_\tau = 322.64$ kt/a, is regarded as the calculation basis of the absolute index of various earnings
④ The price difference of Pb is $V_N - V_w = 4.68$ USD/t and $V_w = 145.94$ USD/t
Source Mao et al. [12]

example, in the Pb circular flow in 1999, compared with a Pb linear flow, when a lead-acid battery with 1 t of lead tolerance is produced each time, 0.321 t of lead concentrate could be saved, and 0.321 t of Pb emission could be reduced via a Pb circular flow, thus increasing the incomes for the Pb production department and the manufacturing sector of lead-acid batteries by 0.274 USD and 26.707 USD, respectively. In a situation with the same resources and environmental load, the output of lead-acid batteries would increase approximately one-quarter to one-third. If a Pb circular flow status in China's lead-acid battery system is improved to the level of the similar system in Sweden between 1992 and 1995 and produces the same amount of lead-acid batteries as in 1999, 359,300 t of lead concentrate could be saved and 3.5929 million tons of Pb emission could be reduced, increasing the income of the Pb production department and the manufacturing sector to 1.67 million USD and 5.82 million USD, respectively. In a situation with the same resources and environmental load, there would be an approximate 320,000 t increase in lead-acid batteries. In the manufacturing stage of lead-acid batteries in Table 8.3, the economic benefit under the prospective condition is lower than the economic benefit in the current situation, although this situation might be caused by the questionable data obtained in China's survey. The major cause of the difference is that during the manufacturing process of lead-acid batteries, the Pb utilization efficiency in Sweden is lower than the corresponding value in China.

The analysis of the economic impact of the material circular flow was to be completed by the STAF team at Yale University. Although professional economists were invited to meet in 2005, because of the obstacles caused by the interdisciplinary in analyzing a material flow, it was difficult to clarify the relationship, and the team failed to achieve the expected research results. However, it is foreseeable in the near future that with the increase in interdisciplinary fields and the support to interdisciplinary fields from the state, more far-sighted economists will provide input into related studies and gratifying research results can be achieved.

In this chapter, the value flow framework based on material flow has been smoothly applied to analyze the economic benefit brought to the enterprise via material circular flow; concurrently, it has been gradually approved by industry and applied to analyze the value flow of other material. For example, it is applied to analyze the value flow of flowing molten iron. See details in the following recommended reading.

Classroom Discussions and Assignments

Topic 1 Discussion

Based on material circular flow, clarify the internal coupling relationship between material flow and energy flow/value flow as the important topic of a circular economy. This was the study to be implemented by the STAF team at Yale University and is the scientific issue to be further researched. Furthermore, from the perspective of natural science funds supported by the state in recent years, an increasing number of people are attempting to perform related research. In this chapter, beginning with the simplest model framework, partial research results are shared. There is a significant number of problems to be further studied. Therefore, please consider and discuss the following topics:

(1) A certain material has multiple purposes; how can you explain its relationship between material flow and value flow?
(2) In the same service or product, there are various substances, and each is important on a large scale, for example, there is M, N in an alloy. Under this circumstance, it is difficult to choose one substance as representative of the product; therefore, how do you analyze the problem of value flow?
(3) What type of coupling relationship exists among material, energy, and value in a circular economy? How do you explain them?
(4) In a circular economy, what type of coupling relationship exists among material flow, energy flow, and value flow among multiple materials in multiple services? How do you explain them?

Topic 2 Discussion

Previously, we have discussed the anthropogenic flow of material and energy and the flow of value. What type of practical significance do these flows have? How do you apply the latter in practice? Please choose a familiar region, and specific to a typical resource or environmental problem, discuss how to explore the solutions to the problems via the above methods. Create a research outline.

Homework and Presentation After Class

Specific to the topic of value flow discussed in class, locate a related paper for class sharing in a group.

Alternatively, specific to a selected topic, locate an application paper of the analysis of material flow, energy flow, and value flow under specific dimensions for sharing in class groups. The grading scale is shown in Table 8.5.

Table 8.5 Grading of the group report in class

Report title	Process of topic selection (maximum 10 points)	Research basis (maximum 10 points)	Research method (maximum 10 points)	Result discussion (maximum 10 points)	Total points

Recommended and Referenced Literatures

1. Zhong, M. 2006. *Introduction to environment and resource economics*, 2nd ed. Beijing: Higher Education Press.
2. Mao, J.S., Z.W. Ying, and Z.W. LU. 2008. The influence of recycling of material on value source intensity. *Proceedings of Information Technology and Environmental System Sciences* 2: 1078–1084.
3. Yan, L., and A.J. Wang. 2014. Based on material flow analysis: Value chain analysis of China iron resources. *Resources, Conservation and Recycling* 91: 52–61.
4. Joosten, L.A.J., M.P. Hekkert, and E. Worrell. 1999. STREAMS: A new method for analysis material flows through society. *Resources, Conservation and Recycling* 27 (3): 249–266.
5. Mao, Jiansu, and Lu, Zhongwu. 2003. The material circular flow and value circular flow. *Journal of Materials and Metallurgy* 2: 157–160.
6. Xilin, Shi. 1983. *Lead accumulator*, 5–6. Beijing: Posts and Telecom Press.
7. Karlsson, S. 1999. Closing the technospheric flows of toxic metals—modeling lead losses from a LAB system for Sweden. *Journal of Industrial Ecology* 3 (1): 23–40.
8. Frosch, R.A. 1994. Closing the loop on waste materials. *The greening of industrial ecosystems*, 37–47. Washington DC: National Academy Press.
9. David, T.A., and B. Nasrin. 1994. *Wastes as raw materials. The greening of industrial ecosystems*, 69–89. Washington DC: National Academy Press.
10. Deacon, R.T. 1998. Research trend and opportunities in environmental and nature resource economics. *Environmental and Resource Economics* 11 (3, 4): 383–397.
11. Turner, R.K., R. Salmons, and J. Powell, et al. 1998. Green taxes,waste management and political economy. *Journal of Environmental Management* 53: 121–136.
12. Mao, J.S., Z.W. Lu, Z.F. Yang. 2004. Several benefits from recycling of industrial materials. *Proceedings of the Fifth Annual Conference for Young Scientists of China Association for Science and Technology* 11 (2–5): 489–498.
13. Marimin, M.A.D., M.P.I.F.P. Machfud, and W. Bangkit. 2014. Value chain analysis for green productivity improvement in the natural rubber supply chain: A case study. *Journal of Cleaner Production* 85 (15): 201–211.
14. Bin, Yu., Fang Zhigeng, Yang Baohua, et al. 2009. A new GERT network model and its application based on value flow process. *Systems Engineering* 27 (7): 43–49.
15. Jiansu, Mao. 2016. *Anthropogenic flow of lead*. Beijing: Science Press.
16. Jiansu, Mao. 2003. *Industrial metabolism of lead and its influence on national economy*. Shenyang: Northeastern University.

9 Implementation of a Circular Economy

9.1 Presentation of Core Issues

A circular economy is an economic form with material circular flow as its core, and it attempts to reduce the influences of human activity on the natural environment system via the flow of closed material in the human socio-economic system, based on phased and efficient utilization of the energy flowing into the human socio-economic system. To achieve this goal and based on the classification of material and energy flows (i.e., they are divided into natural and anthropogenic flows of material and energy based on whether the forming process of the flow is disturbed by human activity), the problems of material and energy flows were discussed in the previous two chapters theoretically, including the anthropogenic flow processes of material and energy, several flow laws, and the possible influences of these flows on the natural flow. How to fulfill and implement the above theoretic flow model in a circular economy? Whether any substance is suitable for recycling? If not, what are appropriate and inappropriate substances for recycling? How can we define the recycling and reproduction of material? What is the pattern of representation? Which problems should be of concern in application? These questions must be explored and discussed by means of practical experience.

This chapter is divided into two sections. The first explores the economic form of a circular economy. What are its characteristics? The second section concerns the methods to promote and implement a circular economy. What type of material is suitable for recycling? How are different recyclable materials recycled? Which problems should be of concern in practice? To answer the question related to the possibility of material recycling, we will define the cyclic regeneration of material, and then analyze how to classify and how do all kinds of material circulate?

J. Mao et al., *Circular Economy and Sustainable Development Enterprises*,
https://doi.org/10.1007/978-981-10-8524-6_9

9.2 What Is the Economic Form of a Circular Economy?

9.2.1 What Is a Circular Economy?

(I) Concept of a circular economy

A circular economy is an economic form with a material circular flow as its core; it guides the socio-economic activity of human beings by using ecological laws. Material resources, energy sources, and wastes are comprehensively utilized to achieve "low mining, high utilization, and low emission," to achieve an economic model in which human activity causes minimum disturbance to the environment, also called an ecological economy.

However, in an ecological economy, the human socio-economic system is typically regarded as the organic component of the earth's surface ecosystem, and it is emphasized that the development of the human social economy should obey ecological priority rules, the disturbance level to the external environment must be controlled within the ecological carrying capacity, and harmonious development between human beings and the environment is required. The coordinating element is related to many relevant factors between human beings and the environment. While the existing circular economy emphasizes material recycling, it is the economic form that emerged to solve the current more significant resource shortages and environmental degradation. The material elements therein are emphatically related; therefore, this is a necessary stage for human beings to move toward an ecological economy and is also the initial stage of the ecological economy implemented under the current environmental conditions.

(II) Comparison between a circular economy and a traditional economy

Compared with a traditional economy, the differences of a circular economy are as follows.

From the perspective of qualitative analysis, in a traditional economy, the material flow is the material's one-way flow formed by "natural resources–product–pollutant discharge." In this type of economy, with the increase in human requirements, additional products are required; therefore, the result is more consumption of natural resources and energy sources, and simultaneously, an increasing amount of wastes and pollutants are discharged into the environment during the production, processing, and consumption processes. If this continues, the result will be the shortage of numerous natural resources and the catastrophic consequence of environmental pollution. Different from this, a circular economy advocates the flow of closed material in the human socio-economic system. The material is organized into the circular flow process of "resources-products-renewable resources" to allow the entire economic system (including production and consumption) to produce no or little resource and environmental load, thus easing the contradiction between human development and the environment.

To further observe the difference between a circular economy and a traditional economy quantitatively, the change relation of the resource efficiency [Eq. (5.7a)] and environmental efficiency [Eq. (5.9a)] obtained from the analysis of material flow are substituted into the governing equation variable form [Eq. (3.3)] after introducing ecological efficiency, respectively, to obtain the relationship of the resource load S with the change in social service quantity S:

$$R = \frac{1 + \gamma - \alpha p}{F_{\mathrm{P}} \cdot \Delta_{\tau}} \cdot S \tag{9.1a}$$

The relationship of the environmental emission load Q with the change in social service quantity is as follows:

$$Q = \frac{\gamma + (1 - \alpha)p}{F_{\mathrm{P}} \cdot \Delta_{\tau}} \cdot S \tag{9.2a}$$

In the above two equations, the meanings of α, γ, p, F_{P}, and τ remain the same as in Chap. 5 and represent the major circulation rate, emission ratio, output change ratio, ratio performance, and service life, respectively.

It can be observed from Eqs. (9.1a) and (9.2a) that the resource and environmental emission loads are related to multiple parameters, such as the variation in social services quantity, major circulation rate, emission ratio, output change ratio, ratio performance, and service life. Under a circular economy, when the circulation rate in the above equation is not 0, there is a nonlinear variation relation between the environmental load and the social services quantity. When material is not being recycled, the value of major circulation is 0 and the above two equations are simplified as follows:

$$R = \frac{1 + \gamma}{F_{\mathrm{P}} \cdot \Delta_{\tau}} \cdot S \tag{9.1b}$$

$$Q = \frac{\gamma + p}{F_{\mathrm{P}} \cdot \Delta_{\tau}} \cdot S \tag{9.2b}$$

Equations (9.1b) and (9.2b) show that the variation relation between environmental load and social services quantity is only related to emission ratio, ratio performance, service life, and other technological production parameters, but it is not related to the circulation rate of material. During a certain period, the technological production parameters remain usually constant, which makes the coefficient of variation between the environmental load and social services quantity also maintain a specific value; thus, the environmental load shows the linear variation relation with the social services quantity, reflecting the variation relation between the environmental load and social services quantity in a traditional economy where material is not recycled.

(III) Basic principle of a circular economy

The implementation of a circular economy should be based on the three basic principles of *reduce*, *reuse*, and *recycle*. "Reduce" requires the related enterprises to reduce material or energy consumption during various of processes such as production, transportation, and uses of products, so as to avoid waste. It includes limited mining of natural resources, implementation of "less material and more yield" resource efficient utilization, and low emission of wastes. For example, simple consumption is advocated to avoid luxury and waste. The daily use of an air conditioner to cool in summer at extremely low temperature and heat in winter at extremely high temperature is a counter-example of reduction during an energy consumption process. "Reuse" requires the repeated use of old objects, such as the recovery of an empty fresh milk bottle that is then cleaned by the fresh milk distribution company for reuse, extension of service life via maintenance, replacement of parts and repair, or the reuse of all or some of a product's parts via refurbishment and remanufacturing, e.g., the replacement of certain components in a computer or the waste residue produced by a power plant being used by a brick company for producing bricks or for paving. Finally, the "recycle" principle refers to waste recovery conducted specifically to the product or part incapable of being reused as a secondary resource for material reproduction, achieving the reuse of material resources. For example, after being recovered, the waste battery in the previous cases will be placed into the production of secondary lead.

(IV) Connotation of a circular economy

It is stipulated by the basic rules of a material cycle that the development of human beings' social economy should obey the natural laws, as embodied in the aspects below.

A circular economy is a new development concept. The human socio-economic system is stressed as the organic component of a natural system instead of replacing the natural ecosystem, as in the traditional economic system, and harmonious development between human beings and the resource environment under the environmental carrying capacity is required. Product transformation is encouraged into service quality enjoyment over ownership, and it is advocated that the "more and fast" size and speed valued in the development of a traditional economy are transformed into the economic development quality valuing a healthy "fine and steady" scientific development concept.

A circular economy is a new working concept. From the perspective of resource types being used, we strive to have a circular economy transform into the use of renewable resources from non-renewable resources and the use of plentiful resources from scarce resources. From the perspective of resource utilization level, the technology and management levels should be strengthened in a circular economy to improve utilization efficiency of resources in multiple areas.

A circular economy is a new model of production. Embodied within different enterprises and based on a "symbiotic relationship," enterprises are built to form an industrial ecological chain, ecosystem industry system, or ecological industrial community. Within each enterprise, cleaner production should be conducted to improve business management and manufacturing technology levels, thus achieving cost savings and emission reductions and increasing the efficiency of the enterprise.

A circular economy is a new economic form. The traditional economic form with a core of production and circulation is expanded into the composite economic form integrated with production, circulation, maintenance, waste recovery, lease, and service.

9.2.2 What Type of Material Is Suitable for Recycling? How Should It Be Recycled?

(I) Type of materials

In practical application, not all materials are suitable for recycling; therefore, material is defined by its "cyclic regeneration," including two concepts of "recoverability" and "renewability." Recoverability means that the product formed from the material can be collected after scrapping, such as glass and paper; this type of product has recoverability. Conversely, products such as insecticide and detergent dissipate into the environment after being used, and they are difficult to recycle; this type of product has no recoverability. Renewability means that wastes maintain their original basic properties after going through regeneration. Put differently, they have the same property as the material that is obtained from natural resources for processing; accordingly, they have the same purpose. In application, the material produced from natural resources is called primordial matter and that produced from waste resources is called a regeneration substance. Additionally, material both having recoverability and renewability is called a cyclic regeneration substance.

What must be noted is that different materials constituting the same product have different cyclic regeneration. For example, as a product, paper has recoverability. As waste paper, there is fiber as the basic composition of the paper and carbon ink from printing. During the regenerative process of waste paper, the paper fiber is recycled, while the carbon ink is discharged into the environment as waste and pollutant; therefore, there is renewability in the paper fiber but not the carbon ink. In application, different types of material should be treated differently.

The cyclic regeneration of material is specific to the existing technology and economic condition. In 2001, based on cyclic regeneration, material was divided into the following three types by Ayres:

Type I The cyclic regeneration of this type of material is feasible based on technology and is economical in financial terms. It includes various metals (structural metallic material is given priority), glass, paper, catalysts, water, and plastic.

Type II The cyclic regeneration of this type of material is feasible based on technology, but it is not always economical in financial terms. It includes some building materials, packing materials, and solvents.

Type III It is difficult to perform cyclic regeneration for these materials, which include surface coating, paint, insecticide, herbicide, preservative, anti-freezing agent, explosive, fuel, detergent, and other chemical products.

As technical and economic conditions will change over time in different locations, for the material incapable of being recycled under existing technical–economical conditions, it might be recycled and regenerated several years later with the development and introduction of technologies; thus, it has cyclic regeneration based on these changes. As an example, for a circuit board that contains various noble metals, there is currently a technical obstacle of material regeneration; however, in the future, with a breakthrough of different metal separation technologies, the multimetal regeneration of circuit boards might be achieved. Therefore, the possible cyclic regeneration may be changed.

The three different materials should be treated with different attitudes while implementing a circular economy. For the first type of material, we will find solutions to allow it to be fully recycled. This type of material is the core matter that should be of concern in the implementation of a circular economy; the enterprise is pleased to collect and regenerate the material. For the second type of material, the government should encourage qualified enterprises to develop applicable techniques while preparing policies. For example, the cyclic regeneration of paper has been generally placed into practice and used in developed countries (many books are clearly identified with a mark indicating they are made with recycled paper), while in our country, although in the summit meeting relevant to the circular economy, most of the paper used is made of raw wood pulp. Currently, there is no recycled paper mill in scale production operated by a paper manufacturer. For the third type of material, we should determine methods to develop substitute products and services to achieve an environmentally friendly society. For example, insecticide is substituted with bioanalysis. We have started to see the use of lithium batteries to substitute for lead-acid batteries at an international level.

(II) Type of material cycle

The core of a circular economy is the circulation of material in the socio-economic system of human beings. Based on the different paths of a material cycle, the circulation of material is divided into three levels by academician Lu Zhongwu.

Small circulation refers to the material cycle within an enterprise. For example, leftover material in the downstream process is recycled back to the upstream process as raw materials for reuse; the gradient utilization of water among different stages of an enterprise (after the enterprise's purification process, the water returns to the production process for reuse); or the cyclic utilization of consumables and by-products during production.

Medium circulation refers to the material cycle among enterprises, for example, wastes of a certain downstream industrial department return to an upstream industrial department as raw material for reuse, or the wastes and surplus energy from a certain industry are transported to another industrial department for processing. For example, several enterprises in a specific industrial park have different requirements for water quality. From the enterprise demanding high water quality, the water flows into the enterprise with a lower water quality requirement for gradient utilization after being discharged. Another example is the waste residue produced and discharged by a specific enterprise in an industrial park is placed into production as the raw material for the enterprise manufacturing composite brick within the park.

Major circulation is specific to the material constituting products. It refers to the process by which a product becomes part of waste recovery and returns to the waste reclamation department for reproduction after its service life and scrapping; the material is reused accordingly.

In the implementation of a circular economy, there are different departments and significant work characteristics relevant to the above three circulations. For small circulation, the production department is the focus of control; the enterprise is not only requested to improve resource utilization technologies but also to document the recovery of leftover materials in each process, to achieve efficient utilization of resources and raw materials. For medium circulation, the focus is on a specific region for integrated management. Based on the relationship of the materials among different enterprises within the region and the different quality requirements of each enterprise, a symbiotic network among enterprises should be created. For major circulation, the primary control is material recovery after product scrapping. Each enterprise, including the materials recovery department in various regions and the waste products recovery department of the manufacturing enterprise, is requested to rationally and timely supervise each process from waste recovery, sorting and dismantling to secondary input of production. For special products, by extending the duties of the producer, the enterprise should not only provide product sales, but should recycle the scrapped product and return it to the enterprise for reproduction or material regeneration.

When a circular economy is implemented on the above three levels, the priority among the three principles of reduction, reuse, and recycle (3R) must be considered. In the enterprise and among enterprises, what should be considered first is reduction, attempt to reduce the waste output for each process in the entire enterprise, thus reducing the consumption of natural resources. Then, consider the circulation problem of these wastes. In product use and scrap, what should be considered first is reuse, an attempt to extend the service life of products and reduce disposable products, and then, consider the circulation problem after products are scrapped. These principles and their priority are very important in China because for many enterprises, the level of reduction and cleaner production has been relatively low, and for many products, there is lack of consideration of reduction and extension of service life.

9.3 What Are the Progression Patterns of a Circular Economy?

The development of a circular economy is embodied in different spatial scales, sectors, and administrative departments. It is primarily to build an industrial ecological park at the urban enterprise level and, on a smaller scale, build an ecosystem industry system and energy gradient utilization system at the social level on an urban scale, creating ecological efficiency control on the scale of region, valley, and country and to build a demonstration zone for a circular economy.

9.3.1 Construction of an Ecological Park

(I) What is an eco-industry park?

Eco-industry parks (EIPs) are an industry conglomeration based on the principles of material cycle or energy flow in an ecosystem. Different plants or enterprises are connected to form a co-existing industry of resource sharing, by-product, or waste. It allows the wastes or by-products of the leading enterprise in the park to be transformed into the raw materials or energy sources of another enterprise, to seek the closed recycling of material resources, gradient utilization of energy, and a minimum amount of environmental waste.

The biggest difference between the industrial symbiosis in an eco-industrial park and the commensalism of species in a natural ecosystem is the difference in causes. The commensalism of species in a natural ecosystem is the result of the natural evolution of species, while industrial symbiosis primarily occurs due to the action of market mechanisms or forms via planning. Usually, the industrial symbiosis system formed via planning is better for the environment.

An ecological park is usually implemented on a small regional scale.

(II) What are the construction modes of an ecological park?

From the perspective of domestic and overseas practices, an eco-industrial park is usually built as one of the models below.

(1) Enterprise-leading type—This type has an original enterprise or several enterprises as the core, and the eco-industry park is established by attracting related enterprises in the industrial ecologic chain, e.g., a classic example is

the Danish Kahlenberg eco-industry park (see examples of ecological parks below). Conversely, guided by the enterprise group, an eco-industry park is established by the internal enterprises in the group based on the principles of ecosystem industry and circular economy, e.g., the eco-industry park built by China's North Shandong Petrochemical Enterprise Group to form three highly related eco-industrial chains, namely sulfuric acid co-producing cement via ammonium phosphate by-product ardealite, seawater "versatile utilities of water," and salt–alkali–electricity co-generation.

(2) Industrial-association type—Within this type, enterprises with high industrial relevancy are integrated via the concept of ecology to fully employ the complementary effect of the park. For example, an industrial association between agriculture and industry is strengthened to promote the agricultural eco-industry park oriented by the sustainable development of industry and agriculture. As the core of cane sugar manufacturing, there are many industrial ecological chains developed in Guigang, Guangxi of China, such as "sugarcane–sugar refinery–alcohol made from waste molasses-compound fertilizer made from alcohol wastewater" and "sugarcane–sugar refinery–bagasse pulping-recovery of pulping black liquor alkali"; the eco-industry park is established based on this model.

(3) Reforming and rebuilding type—This type is reformed and rebuilt on the basis of an original industrial park and a high-tech park to create an upgraded eco-industry park for ecological enterprises to gather.

(III) Example of ecological parks

The Danish Kahlenberg eco-industry park is an example and is shown in Fig. 9.1. Focusing on several core enterprises in Kahlenberg, including the existing thermal power plant, oil refinery, plasterboard plant and pharmaceutical factory, and by means of the exchange of substances or energy gradient utilization, the ecological park performs full use of coal, coal ash, sulfur and other resources, gradient utilization of steam, waste heat, and other energy sources at the level of the city of Kahlenberg; therefore, it becomes a classic case of an ecological park.

9.3.2 Construction of an Ecosystem Industry System

(I) Planning method for an ecosystem industry system

Ecological planning for an industry system is an important method to achieve regional industry ecology and is also the foundation of practice.

The planning of an ecosystem industry system is the technological means, based on the principle of material circulation in ecology, and combined with the material flow status of an industry system in a specific area to built industrial network system with a core of competitive or leading industries and a collection of upstream–downstream industries, and to promote the evolution of a regional industry from a material flow to an ideal condition of closed circulation.

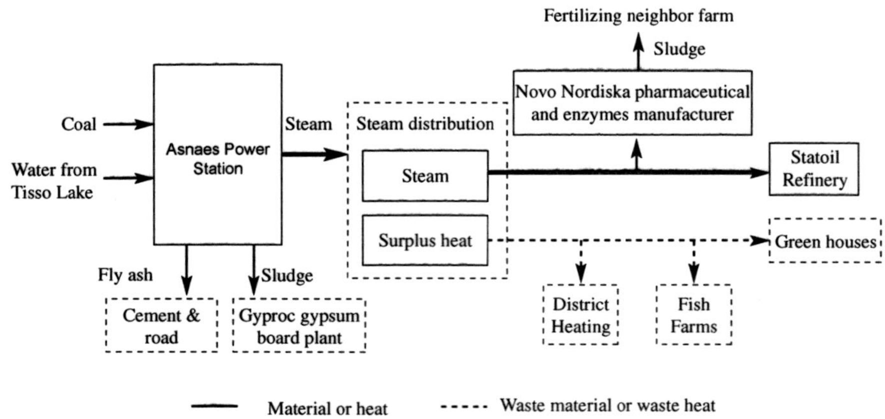

Fig. 9.1 Resources flow of the eco-industrial system of Kahlenberg. Prepared according to the free encyclopedia of Kalundborg eco-industrial park. https://en.wikipedia.org/wiki/Kalundborg_Eco-industrial_Park

During the planning of an ecosystem industry system, first, research the industry system of the region and understand the local competitive or leading industries, the major material source, the material flow process of these industries, their industrial technical level, the local resource support capacity, and the environment capacity of the utilized resources. Second, analyze the flow relationship of the material in different life cycle phases and among different industrial subsystems within the entire industry system, including flow direction, amount distribution, shape transformation, and spatial transference to identify the disharmonious links (e.g., the breaking point among enterprises during material flow) and factors (e.g., inconsistent quantity). Third, a new industry type is introduced to compensate for the defects of the eco-industrial network and improve the degree of association among regional industries. The production scale of an industry is properly adjusted to promote the reasonable distribution of resource utilization. Product reuse is promoted by means of product maintenance and methods for maintenance to extend the product's service life and to reduce the consumption intensity of materials within the industrial system. Dismantling of scrapped products and recovery of parts reduce the consumption of materials, and waste recycling reduces the amount of waste emission and consumption of natural resources. "Less input, high output, and low emission" are achieved, and they produce an integrated effect that is coordinated with external resources and the environmental system. Finally, under the established framework of an ecosystem industry system, system management is implemented for the primary materials and products of the competitive industries, including design for environment (DfE), product development and function substitute, green manufacturing, extended producer responsibility (EPR), green packaging, transporting and recovery, and recycling after product scrapping. As products belong to different administrative departments during different life cycle phases, the planning-based eco-industry system must reintegrate the management system and promote comprehensive management.

Compared with an industrial ecological park, an ecosystem industry system is always implemented at a regional level with a slightly larger scale than an eco-industrial park, as it requires the industrial symbiosis formed via multiple different enterprises; it is available to execute the construction on an urban scale or the scale of a specific functional area.

(II) Example of an ecosystem industry system

The Xiamen eco-industry planning performed between 2004 and 2005 is used as an example.

Through an early survey, it was found that the electrocommunication industry is one of the pillar industries in Xiamen city. In the construction of the electronics and information service ecosystem industry system (based on Dell (China) Co., Ltd. and other original electronic industries combined with the information service industry), the maintenance service functions of the electronic industry should provide extended service periods for its products and supplement the scrap recycling-disposal industry for electronic products to jointly form a "semi-life cycle" industry system of electronic products. This system is shown in Fig. 9.2.

Figure 9.2 shows that the perfect electronic and information service eco-industry system should include electronic component supply, product production, product use, scrap, recycling, and other life cycle phases. In the figure, the perfect degree of the existing industry is represented by the depth of the color; the darker the color is, the better the development of a related industrial sector. This provides an important basis for the next proposal of construction and management measures.

If it is further refined based on the category of electronic products, Fig. 9.2 can be divided into a computer and network product chain, a communication and other information product chain, and a digital product and "all-in-one" household smart center product chain.

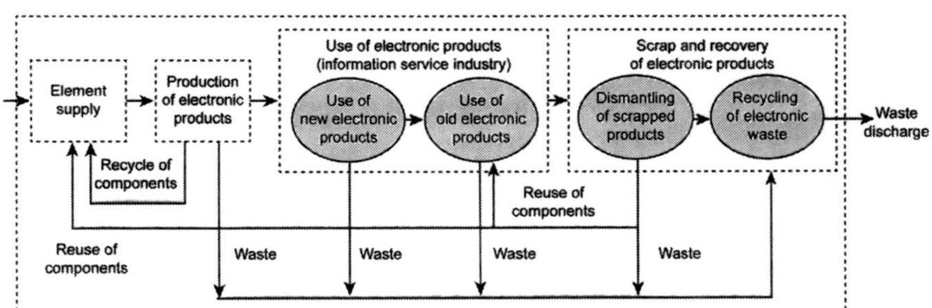

Fig. 9.2 Conceptual model of an electronic and information service eco-industry system 2006. *Source* Mao [21]

9.3.3 Gradient Utilization of Energy Sources

(I) Gradient utilization model of energy sources

Energy gradient utilization is a method to rationally utilize energy sources. Based on the grade of the energy sources, the energy sources are used with different methods in different grades.

For example, in a combined heat and power (CHP) system, high to medium temperature steam is used for electricity generation, or it is used in production that requires high-temperature steam, and low-temperature waste heat is used for heating residences. The grade of energy sources refers to the level that energy sources can be transformed into mechanical work. High-grade energy sources include electric power, fuel gas, and liquid fuel, and low-grade sources primarily include heat energy and bioenergy. Generally, it is simple to transform high-grade energy into low-grade energy, and it has a high energy conversion efficiency; however, it is difficult to transform low-grade energy sources into high-grade energy because a specific technique is required and there is more energy loss in the transformation process. The energy gradient utilization can improve the energy utilization efficiency of the entire system; it is an important measure for energy saving, and it is the main content specific to the development of a circular economy via energy use.

Similar to the recycling of material, energy gradient utilization can be implemented at different levels, such as enterprise or society. In an enterprise, generally, it is guided by producing energy based on the energy-use requirements for each of the different product production processes. The flow of energy gradient utilization is planned and designed to efficiently use the energy sources in the enterprise. From a social perspective and based on the energy-use requirements of the local areas, the requirements of different energy users from the aspects of production and living vary based on energy type, grade, quantity, area of energy use, design, and planning mode. The process of energy gradient utilization and the energy delivery method in this type of area efficiently use energy sources in their specific region. Figure 9.3 is a typical example of a model of energy gradient utilization.

Usually, energy gradient utilization is implemented on a regional level that is larger than an eco-industrial park, even the ecosystem industry system; e.g., it is implemented on an urban scale or from a social perspective.

(II) Example of the basic forms of common energy gradient utilization

There are various forms of application of energy gradient utilization, and they are described by using CHP and combined circulation generation as examples.

Co-generation of heat and power (or CHP generation) combines heat supply and electricity generation within the same power plant. This term is specific to traditional thermal power generation. Usually, thermal power generation only generates one type of product, electricity, and its efficiency is approximately 35%. This means that approximately 2 MJ of heat is wasted for every 1 MJ of electric energy output.

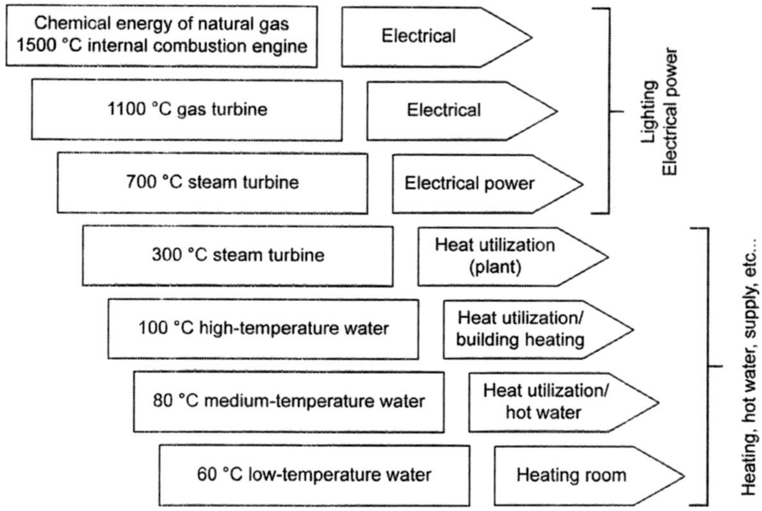

Fig. 9.3 Example of the mode for energy gradient utilization

If this portion of heat is used to heat water, it is available to completely satisfy the demands for heating and hot water for bathing in homes and other plants near the power plant. Generally, a generator is driven by a steam turbine to generate electricity in a CHP process, and the exhaust gas is used to heat the existing boiler plant; its overall efficiency can reach 80%. It can be seen that in the CHP process, the quantity of heat originally discarded by an ordinary power plant is utilized to provide inexpensive heat for warming to industries and families, greatly improving energy-use efficiency. The CHP process not only generates electricity but also produces heat, and compared with separate generation of heat and power, it has the following advantages: reduce energy consumption, reduce pollutant discharge, reduce power consumption via waste-heat utilization, save land area, improve quality of heat supply, facilitate the integrated use of natural resources, improve city image, and reduce accidents. Because of its many advantages, the CHP process has been developed throughout the world.

Combined cycle power generation (CCPG) is specific to the generator set used in the electric production process. It refers to the power generation system integrated with a gas turbine circulation system, heat recovery steam generator (HRSG), and steam turbine circulation system. It transforms the high-temperature exhaust flue gas discharged by the gas turbine into steam via the recovery of HRSG and routes it to the steam turbine for power generation; the steam remaining after power generation is used to supply heat. The common form includes the single-shaft combined cycle, in which a set of generators is driven by a gas turbine and steam turbine coaxially, and the multishaft combined cycle, gas turbine, and steam turbine are combined with the generator. It is primarily used for power generation and the CHP process. The combined cycle unit of a gas turbine has the following unique advantages: ① High generating efficiency—CCPG efficiency can approach 57–58%, much higher than an ordinary coal-fired power plant (with a generating efficiency of 0.75–1000 MW or a range of 20–48%). ② Less pollutant discharge—There are more air pollutants, e.g., dust, SO_2, and nitric oxide (concentration is up to 200×10^{-6}), discharged from the boilers of a coal-fired power plant, arranging desulfurization, denitration, electrostatic precipitation, and other end treatment equipment is necessary. But there is very little dust and little SO_2 in the exhaust gas discharged from the HRSG of a combined cycle power plant, and the concentration of nitric oxide is only 10×10^{-6} to 25×10^{-6}. ③ Flexible mode of operation—A coal-fired power plant has a long start-up time and is suitable for base load operations and has poor peak performance. A gas turbine power plant cannot only perform as a basic load but also as a variable load plant. Various types of fuel (e.g., oil, natural gas, and coal) are used as the source of energy for a CCPG; therefore, it is easy to regulate the peak based on the types of energy used. ④ Less consumption of water—In a gas-steam combined cycle power plant, the power generation of a steam turbine occupies one-third of the capacity; therefore, the water consumption is generally one-third of that of a coal-fired thermal power. In addition, the H in (CH_4) of natural gas and O in the air can be converted into CO_2 and H_2O via combustion. Theoretically, approximately 1.53 kg of water can be recycled by every combustion of 1 m^3 of natural gas. This amount of water is sufficient to satisfy the power plant's water demand. ⑤ Less land area—If there is no storage of coal and ash and an air cooling system can be utilized, the land area is only 10–30% of that of a coal-fired power plant, which greatly saves land resources. ⑥ Short construction cycle—There is a significant difference in the construction cycle of different types of power plants. Generally, a gas turbine plant requires eight to ten months, and a combined cycle system requires 16–20 months, while a coal-fired power plant requires 24–36 months. In addition, a combined cycle system can be used for solar thermal electricity and nuclear power plants.

In short, energy gradient utilization can greatly improve the utilization efficiency of energy sources, not only reducing energy consumption to effectively protect natural resources, but reducing pollutant discharge to effectively improve environment quality, making it important to implement a circular economy.

9.3.4 Management of Ecological Efficiency

(I) Review of the basic laws of ecological efficiency

In Chap. 3, the concept of ecological efficiency was introduced and the relationship between environmental load and ecological efficiency was explored to obtain the status of improved resources and the environment, indicating that ecological efficiency must be greatly improved. The improvement of ecological efficiency means that obtaining the same service quantity will consume fewer natural resources or discharge less environmental wastes and pollutants.

In Chap. 5, specific to a single product system, the change rule of an ecosystem was explored to determine that the ecological efficiency is related to the major circulation rate α of material, output change ratio p, product service life $\Delta\tau$, material emission ratio γ, and product ratio performance F_P. The improvement of a material's circulation rate is beneficial to improve the ecological efficiency.

In the case analysis in Chap. 7, the source of energy is used as an example to analyze the basic relationship between the ecological efficiency of an industrial system and its constituting industries to determine that the improvement of the ecological efficiency of the entire industrial system must not only improve the energy efficiency of each industry but also adjust the industry structure to improve the economic segments of the sector with higher energy efficiency.

In an actual economic system, the material or energy flows among many enterprises are neither the pure "series-type" structure nor the pure "in parallel" structure, but are a mixture and a complex network structure. The ecological efficiency of the entire system is closely related to its composition and the ecological efficiency of each component; therefore, we must not only adjust and optimize the system structure, but we must promote the ecological efficiency of each subsystem or component.

Management of ecological efficiency can be implemented from multiple perspectives and levels; for example, a study of a specific material, product system, industry system or region is an important method to obtain a scientific management method and improvement strategy.

(II) Example of the changes in the ecological efficiency of an industrial sector

Industry energy consumption is used as an example.

To understand the distribution characteristics of industrial energy efficiency in different regions, several cities with available data are selected to calculate the industrial distribution of industry industrial sectors in different regions based on the method in the case analysis in Chap. 7, providing a basis for the management of a regional industrial sector. Figure 9.4 shows the energy efficiency of an industrial sector in three cities (Chongqing, Tianjin, and Guangzhou) in China (exclusive of Hong Kong, Macao, and Taiwan data) in 2007. It shows the variation of our country's energy efficiency with the change of industrial sector and region.

It can be observed from Fig. 9.4 that in the cities of Guangzhou, Chongqing, and Tianjin, the industrial sectors whose energy efficiency is significant (exceeding 200,000 Yuan/tec) are Guangzhou's tobacco industry (code TOM) and Chongqing's furniture manufacturing (code FNM), stationery and sporting goods manufacturing (code ARM), special equipment manufacturing (code SMM), and gas production and supply (code GPS) (from left to right). However, the energy efficiency of various sectors in Tianjin is below 100,000 yuan/tce. Among the various industrial sectors in China, the energy efficiency of the tobacco industry (code TOM) is the only one to exceed 100,000 yuan/tce; the energy efficiency of other industries is below 100,000 yuan/tce. The results show that for different cities or regions, the maximum energy efficiency comes from the different industrial sectors;

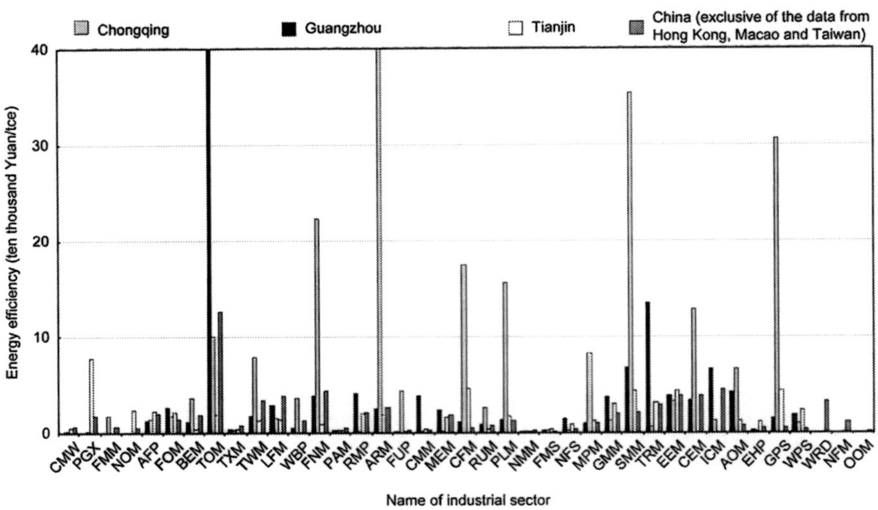

Fig. 9.4 Comparison of the energy efficiency of industrial sectors among different regions. Reprinted from Mao et al. [14], the industrial code is same as the Table 7.2. Based on GB/T 4754-2002

furthermore, there is a significant difference in the energy efficiency of an industrial sector from several tens to hundreds of times. The energy efficiency in the same industrial sector obviously varies based on different cities or regions, which shows the imbalanced energy efficiency of an industrial sector in different areas. The imbalance of the energy efficiency of an industrial sector means that there is enormous potential for improvement of the industry's energy efficiency.

9.3.5 Demonstration Area of a Circular Economy

(I) Summary of a demonstration area of a circular economy

A circular economy demonstration area (hereinafter referred to as the demonstration area) is a demonstration region with pollution prevention as its starting point and material circular flow as its characteristic; its final objectives are social, economic, and environmental sustainable development. The socio-economic activities within regions are organized into several feedback processes of "resources-products-renewable resources" by using ecological laws to attempt to control the production of wastes during production and consumption. The goal is to recycle available products and wastes and rationally treat and dispose products incapable of being used to achieve "low mining, high utilization, and low emission" of material production and consumption, thus effectively utilizing resources and energy sources, reducing pollutant discharge and promoting harmonious development between the environment and the economy.

Compared with an eco-industrial park, a circular economy demonstration area contains complex content and has a broader scope. It is a type of bio-economic zone within a regional sphere that not only establishes an industry network among enterprises but also combines the material cycle and energy flow among the primary, secondary, and tertiary industries, therefore achieving the overall circulation of the industry within a regional economy. Adjustment of macroeconomic policy and reestablishment of legal institutions will provide strong guarantees for the demonstration area of a circular economy.

The main concept of a circular economy's demonstration area is to combine the development of a circular economy with the comparative advantages of regions and focus on ecological protection and environmental governance. Through the introduction of high and new technology, remolding the existing unreasonable regional industry and building the production and consumption patterns for a circular economy allow it to conform to the local specific advantages.

During the construction of a circular economy demonstration area, the 3R principle (reduce, reuse, and recycle) is followed. The 3R principle is the core of the theory of a circular economy and must be followed without exception during the construction of a circular economy demonstration area. Reduction of input, reuse during the flow, and recycle of global resources and wastes are indispensable.

(II) Steps in the construction of a circular economy demonstration area

The construction of a circular economy demonstration area in our country is divided into four steps: planning of the demonstration area, hardware construction, software support, and construction of a management index system.

For the demonstration area's planning, first, build a planning team that contains a leader and a technical organization; then, investigate the existing situation to analyze the local natural conditions, social background, ecological environment, and the existing economic operation model. Next, the building objectives and specific design are determined, including the overall framework design, industrial development plan, ecological landscape plan, major project screening, and preparation of laws and regulations. Finally, an analysis of the investment and benefits is performed, primarily referring to the investment budget for the construction and analysis of social, economic, and environmental benefits.

Hardware construction is primarily focused on the three levels of enterprise, region, and society. At the enterprise level, the goals are to achieve cleaner production and minimize sewage disposal, improve techniques and energy saving and emission reduction, and develop and utilize the waste resources produced in the enterprise. At the regional level, the theories of ecology and circular economy are applied to classify an existing industrial park for guidance. An eco-industry park is built to promote the grade and competitiveness of the existing economical–technological development area, and a new high-tech development zone is developed to guide the transformation of old industrial areas. Especially in areas where resources have been exhausted, economic transition should be accelerated. At the social level, a resource recycling society must be developed, including a classification, recovery, and recycling system for municipal solid and other wastes, a city and regional reclaimed water reuse system, an ecotype industry system and an information system. The primary, secondary, and tertiary industries are considered in the aggregate to uniformly plan the material cycles and energy flows in and among the three industries.

For software construction, there are two aspects of laws and regulations to support system and technical support system design. Laws and regulations provide policy guarantee to the construction of a demonstration area, including the construction of the development system for the circular economy laws and regulations, preparation of the preferential policy for circular economy development, and encouragement for green consumption and procurement. The technical support system primarily includes environmental engineering technology, waste resource utilization technology, and cleaner production techniques. During the construction of a circular economy demonstration area, focus must be on the development and application of the key connection technologies of an ecological industry, to connect the units of the entire demonstration area and to apply advanced technologies to practical construction.

The indicator system of a circular economy demonstration area includes four types: indicator of economic development, characteristic index of circular economy, indicator of ecological environmental protection, and indicator of green management. Using the local GDP as the basis, the indicator of economic development is used to evaluate the local economic growth. The characteristic index of a circular economy includes resource productivity, the cyclic utilization rate, and final handling capacity, representing the input to the demonstration area, the productive process, and the emission load, respectively. The indicator of ecological environmental protection refers to the environmental quality, environmental performance, ecological construction and improvement potential of the ecological environment and other environmental conditions. The indicator of green management refers to the policies and rules management indicator to promote the construction of a circular economy and the indicator of an enterprise's management and public awareness.

Classroom Discussions and Assignments

Topic Discussion

A circular economy involves each field of a social economy; therefore, everyone must participate and assume responsibility for their corresponding field. Please consider and discuss the following topics:

Table 9.1 Grading of the group report in class

Report title	Process of topic selection (maximum 10 points)	Research basis (maximum 10 points)	Research method (maximum 10 points)	Result discussion (maximum 10 points)	Total points

What can we do to implement a circular economy?

Based on a typical problem in a hometown or specific region, discuss how to implement a circular economy.

Homework and Presentation in Class

Specific to the topics discussed in class, locate research papers concerning cases of circular economy for class discussion via the group. The grading scale is shown in Table 9.1.

Recommended and Referenced Literatures

1. Graedel, T.E., and B.R. Allenby. 2003. *Industrial ecology*, 2nd ed. Prentice Hall.
2. Ayres, R.U. 2001. *Industrial metabolism:Theory and policy,the greening of industrial ecosystems*, 23–37. Washington D C: National Academy Press.
3. Ayres, R.U. 1994. *Industrial metabolism:Theory and policy: The greening of industrial ecosystems*, 23–27. Washington DC: National Academy Press.
4. Jianqiang, Bao, and Huang Haifeng. 2009. *Outline of circular economy*. Beijing: Science Press.
5. Nansheng, Deng. 2002. *Industrial ecology-theory and application*. Beijing: Chemical Industry Press.
6. Aile, Lao, and Geng Yong. 2003. *Industrial ecology and eco-industry park*. Beijing: Chemical Industry Press.
7. Jingjiang, Li. 2006. *Enterprise' green operation-the only way to sustainable development*. Beijing: Tsinghua University Press.
8. Lu, Zhongwu. 2003. Study on some aspects of circular economy. *Research of Environmental Sciences* 16 (5): 1–5, 10.
9. Zhongwu, Lu. 2009. *Basis of Industrial Ecology*. Beijing: Science Press.
10. Mao, Jiansu, Zeng Run, and Du Yanchun, et al. 2010. Eco-efficiency of industry sectors for China. *Environmental Science* 31 (12): 269–275.
11. Wang, Lihong. 2005. *Circular economy: Ways to implement the strategy of sustainable development*. Beijing: China Environmental Science Press, 张.
12. Yang, Zhang. 2005. *Outline of circular economy*. Changsha: Hunan People's Publishing House.
13. Zhao Tao, and Xu Fengjun. 2008. *Outline of circular economy*. Tianjin: Tianjin University Press.
14. Mao, J.S., Y.C. DU, and H. CAI, et al. 2010. Energy efficiencies of industrial sectors for China' s major cities. *Procedia Environmental Sciences* (2): 781–791.
15. Mao, J.S. 2012. *Planning of industrial system*//YANG Z F. Eco-cities: A planning guide. CRC Press.
16. Zhongwu, Lu. 2008. *Study on industrial ecology-across "environmental mountain"*. Beijing: Science Press.
17. Zhongwu, Lu. 2009. *Basis of industrial ecology*. Beijing: Science Press.

18. Yi, Qian. 2006. *Cleaner production and circular economy: Concept, method and case*. Beijing: Tsinghua University Press.
19. Graedel T.E., and B.R. Allenby. 2003. *Industrial ecology*, trans. H. Shi. Beijing: Tsinghua University Press.
20. State Environmental Protection Administration. 2010. *Planning guide for circular economy demonstration area (Trial) [EB/OL]*. [2010-08-09] http://www.mep.gov.cn/info/gw/huanfa/200312/t20031231_87885.htm.
21. Mao Jiansu. 2006. *Study on urban eco-industry planning*. Postdoctoral Research Working Report of Beijing Normal University. Beijing: Beijing Normal University.

10 Circular Economy Guarantee System

10.1 Presentation of Critical Issues

In the previous chapters, we defined the pattern of a circular economy, including the type of flow relationship that material energy and value should have and the type of implementation method to apply to different scales ranging from enterprise to region; all of these become important objectives when implementing a circular economy. However, if there is no "channel" to achieve this goal, it is difficult to guarantee that the material, energy, value, and other relevant factors can achieve the objectives in the human socio-economic system. Therefore, the preparation of the guarantee system becomes an important factor when implementing a circular economy. The method to guarantee the implementation of a circular economy is the key topic of this chapter.

To determine a method to guarantee the implementation of a circular economy, we not only must clearly identify who is responsible for the implementation of the circular economy and how it can be established, but we must clearly identify the reason. Thus, we must discuss and clearly identify the influential factors in a circular economy, identify the inner relations among these factors, and determine how they influence a circular economy. On this basis, an organic integrity composed of various influential factors is established to assist in the implementation of the stated objective, specifically, the circular economy; the organic integrity consists of various influential factors. First, we clarify the concept of guarantee in this chapter and identify the factors involved in the implementation of a circular economy and their relationships, so that we can clarify the meaning of a guarantee system and establish the structure of a circular economy guarantee system. Second, we clarify how to guarantee the implementation of a circular economy. Overseas examples and our country's circular economy laws will be presented for discussion and analysis.

J. Mao et al., *Circular Economy and Sustainable Development Enterprises*,
https://doi.org/10.1007/978-981-10-8524-6_10

10.2 What Type of Guarantee System Should a Circular Economy Have?

10.2.1 What Is a Guarantee System?

(1) Concept of a guarantee system

In the Chinese dictionary, "guarantee" has the following three meanings: ① to keep (life, property, and right) from infringement, e.g., a guarantee of national security; ② to ensure and guarantee fulfillment, e.g., guarantee of supply; and ③ the thing or object playing a guarantee role. Usually, "guarantee" includes three key elements: The first is the security objective, to clarify who or what should be guaranteed, e.g., a specific person or thing. Second is the job objective, the expected specific function, purpose, or condition. Third is the implementation method; put differently, what type of measures should be employed to ensure the implementation of the previous objective? It is not difficult to observe that many factors would be involved in the implementation process of the entire event to achieve a specific objective; furthermore, there is an inherent connection among elements that will develop into the integrity with specific functions formed via interrelations and mutual dynamic support among the various related things or elements—this can be called the guarantee system.

Several examples are provided to assist in understanding.

Example 1. A traffic guarantee system is required for transported cargo to ensure that it arrives at the destination. Here, the security object is the cargo to be transmitted, the job objective is to ensure that the cargo is free from any damage and is not sent to another location, and the implementation method may be shipping via a specific logistics company. In practical implementation, to ensure the above objective, manpower must be organized to understand the basic information regarding the cargo and perform necessary packaging. It further demands the connection of cargo receiving, transportation, examination, and other processes along with the preparation, determination, and implementation of necessary measures in each process. It can be observed that the process involves various elements, and there exists a specific inner relation between each element and the implementation of a system objective; thus, it has organic integrity.

Example 2. The heating ventilation air-conditioning system has a goal of meeting the specific requirements of temperature, humidity, and cleanliness. The security object is a room of a specific size, and the job objective is the specific temperature, humidity, and cleanliness of the room, e.g., 27 °C ± 0.5 °C, 65%, and hundred-class clean workshop. The implementation method is to employ the central air-conditioning clarification system. It specifically includes refrigeration, heating supply, cold–hot exchange, humidification–dehumidification, filtration purification, air supply/return pipeline, energy and power supply and information processing and control systems. Each system should be provided with corresponding equipment, pipeline channels connected with each equipment room and other various facilities and the corresponding managers to form organic integrity integrated with various manpower and material resources.

(II) Examples of a guarantee system

There are different work objects and job objectives in different fields; accordingly, there are different guarantee systems. The social security system and safety guarantee system are used as examples of specifications.

1. Social security system

The security object of the social security system is society, and the job objective is social stability and sound development. Its guarantee system is the integrity formed via the interrelation and mutual support of each component relevant to the security of healthy social development. In terms of our country and specific to different groups of people, it is divided into three layers of guarantee. The first is the "bottom layer." Social assistance is offered to the poor urban and rural families and residents mainly via subsistence allowances, medical assistance, rural "five guarantees," and other systems, to support widowers and widows and other specific groups via the social welfare system. The required funds are completely provided by the government. The second is the "trunk layer." It offers provision for the aged, medical treatment, unemployment, occupational injury, fertility, and other basic guarantees for the insured primarily via the implementation of the social security system relevant to rights and obligations. The government forces or guides the implementation of the social security system and bears the necessary fiscal responsibility. The third is the "supplementary layer." Based on conditions, an organization is organized and guided by the country to establish enterprise and occupational annuities, a supplemental medical insurance system and commercial insurance to meet the guarantee needs of different members of society. Overall, by means of social insurance, social relief, social welfare, preferential treatment, and other measures, our country protects social stability, promotes social equality, and satisfies the basic living needs of rural and urban residents; supports the sound development of society and the economy; and ensures the reproduction of a social labor force. The social security system is the safety net of society and is critical to social stability and development.

2. Emergency safety guarantee system for hazardous articles

An accident emergency rescue safety guarantee system for dangerous chemicals refers to a system for emergency rescue from dangerous chemicals. The security object is the rescue team, and when the evacuation of trapped human beings and the affected area is successfully completed, the rescue team can be protected to the maximum extent. The realization of the primary guarantee objective in this system requires that the rescuers with special devices possess good situational awareness and team spirit and that the hardware and software that provide information are efficient so that the system used by the rescuers can be improved under the guidance of a peacetime union concept.

The structural support of the system is shown in Fig. 10.1. In the figure, the dangerous chemicals emergency management platform is closely collaborative with the dangerous chemicals security supervision network; both are entities of the state crisis management system, and their development depends on the improvement of modern macro-management concepts of state crisis management. The emergency rescue force addressing the dangerous chemicals belongs to the dangerous chemical emergency management entity and the security supervision network for dangerous chemicals in the information sharing space. The system at the theoretical level requires the guarantee of various hardware and software. In the bottom layer of the figure, command decision support, individual protection support, communication guarantee support, think tank technique support, legal services support, risk study and judgment support and psychological service support are related entities that

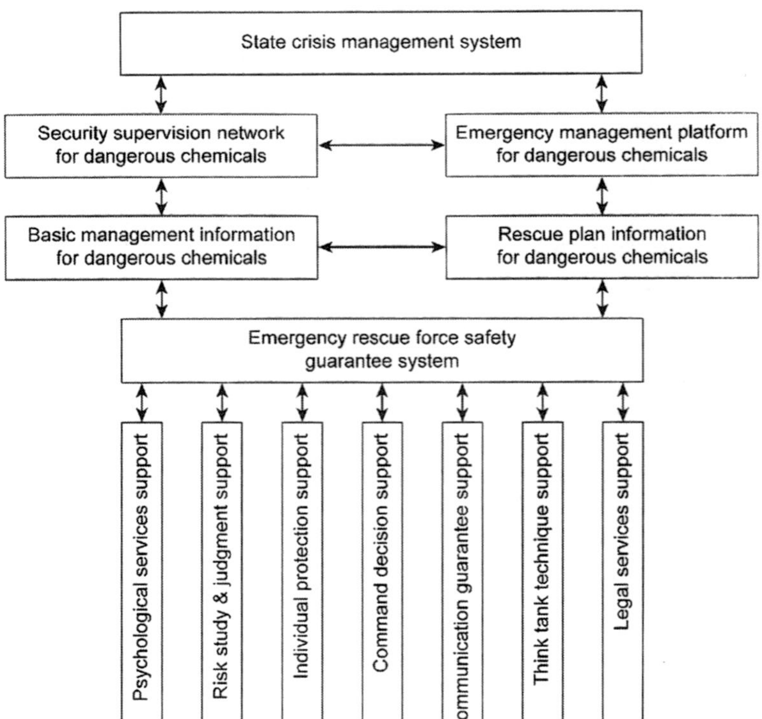

Fig. 10.1 Support structure of the dangerous chemicals emergency rescue guarantee system. *Source* Li et al. [10]

constitute the accident emergency rescue from dangerous chemicals. In the seven support structures, the communication guarantee support, individual protection support, think tank technique support and risk study and judgment support focus on decision-making support and support the formation of good situational awareness; they are the introduction point for the system operation. As the connecting points of commend system to operation system, the risk study and judgment support, think tank technique support and the outermost layer of legal services support and psychological service support are critical for system operation.

(III) Implementation elements of a circular economy

In the circular economy guarantee system, the object is the various materials and energy demanded by the sustainable development of the economy, and the job objective of the guarantee is to satisfy the material needs of human beings with specific location and time distribution characteristics, namely to place raw materials into production from the human socio-economic system, reuse all wastes capable of being recycled and regenerated by human beings, and identify a substance capable of forming the full-closed circular flow in the human socio-economic system based on Fig. 5.5. The energy management objectives include utilizing renewable energy sources as much as possible and optimizing the energy system so that various sources of energy can be most adequately and efficiently utilized based on the gradient utilization mode of energy sources shown in Fig. 9.3. As the transition stage to achieve the above ideal scenario, we should attempt to facilitate a semi-closed material flow mode of materials according to Figs. 5.1 or 6.1. Based on the analysis in Chaps. 5 and 6, it is not difficult to observe that the method to achieve such an ambitious goal will be influenced by many factors, such as the type of material resources, production technology, human requirements, and consumption model in the circular economy guarantee system. It must organically integrate related factors based on their inner relations to form a management system capable of ensuring these materials and energy are performing according to the preset model. It must clarify the influential factors of the circular economy and how these factors influence the circular economy. Only by analyzing these relationships can we establish a scientific and effective circular economy guarantee system. Therefore, the guarantee system mentioned in this course refers to the management system to ensure the implementation of a circular economy.

10.2.2 What Are the Influential Factors in a Circular Economy?

As mentioned above, the circular economy is influenced by many factors. A circular economy is interdiscipline and involves multiple fields of study, and different fields and departments bear different tasks; accordingly, the guarantee systems we have experienced are also different. However, this is a new branch of science, so what we have known or experienced is very limited; thus, it is difficult to include all elements related to a circular economy. To understand the influential factors of a circular economy and how they influence the circular economy, we provide several examples and specify the possible influential factors and their influence on the circular economy.

(I) Example 1. Theoretical analysis of the various flows relevant to a trend process of resource service driven by human requirements

The core service of economic activity is to meet specific needs of human beings. Here, the service generation process of mineral resources is used as an example to analyze the relevant flows and to identify several internal influencing factors of the circular economy.

The service trend of mineral resources refers to the realization process of the metal (e.g., Fe and Al) and the energy source (e.g., coal, petroleum, and natural gas) from the lithosphere to providing the specific final services, also known as the service trend process. The process includes resource access, material processing, product manufacturing and use, and other basic activities, as shown in Fig. 10.2. The use process of products refers to material providing a final "service" for human beings, while other processes are the forming process of a material's final service. For physical property, the service trend process not only includes the space variation process of material (e.g., underground crude oil is excavated and processed into gasoline at the Daqing Oil Field and then sent to Harbin and other gas stations

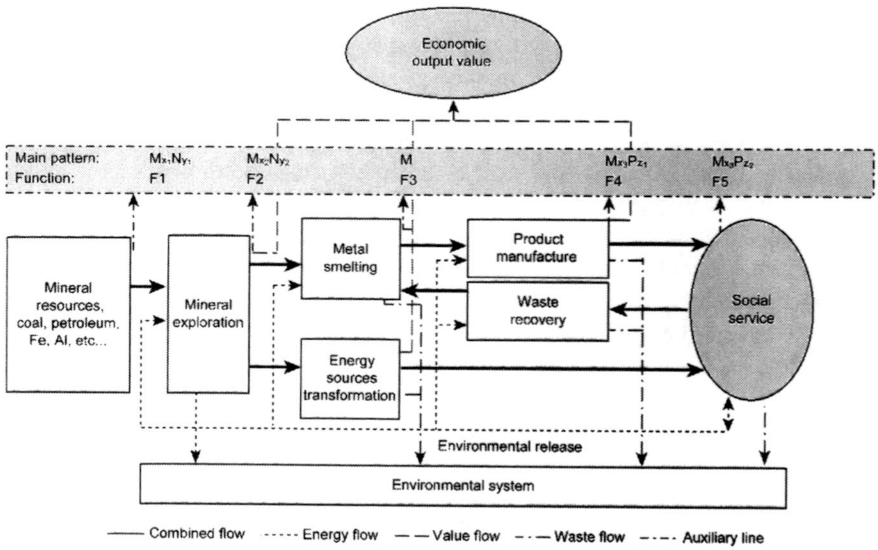

Fig. 10.2 Conceptual framework of the service trend process of mineral resources M—specific metallic element, N, P—other material elements, x, y, z—composition variables, F—substance function. Reprinted from [27], with permission from Environmental Science & Technology (Chinese)

through conveying pipes), but it includes the technical change process of the material (e.g., crude oil is refined into gasoline, natural gas, and other industrial products via cracking) during which a series of technological changes will occur in the material's function and shape.

In the service trend process of mineral resources, for the implementation of material function, shape, and other basic changes, the basic conditions of these change occurrences must be provided, specifically, the corresponding energy, power and manpower, manufacturing technique, equipment and other infrastructure, and the exchange of various materials and sources of energy among different production and user departments, to integrate energy and value flows into the service trend flow process of the material. As material is both the energy and the value carrier, the law of conservation of mass should be obeyed in each material flow, which makes the "material" the link to connect the internal relations among the material, energy, and value flows, to form the specific coupling and synergic relationship among different properties and the quality of the material, energy, and value flows during the service trend process of mineral resources. In addition, the metabolic activity will discharge numerous wastes and waste heat to the external environment. In general, the service trend process of mineral resources is a multifactor, multilevel, multicomponent, and multiscale complex change process. It can be described as the relationship frame of "one point (service target point), two axes (technique-dimension axis of the trend process and time-dimension axis), three levels (inner, middle, and outer layers), and four varieties (material, energy, value, and time)," as shown in Fig. 10.3.

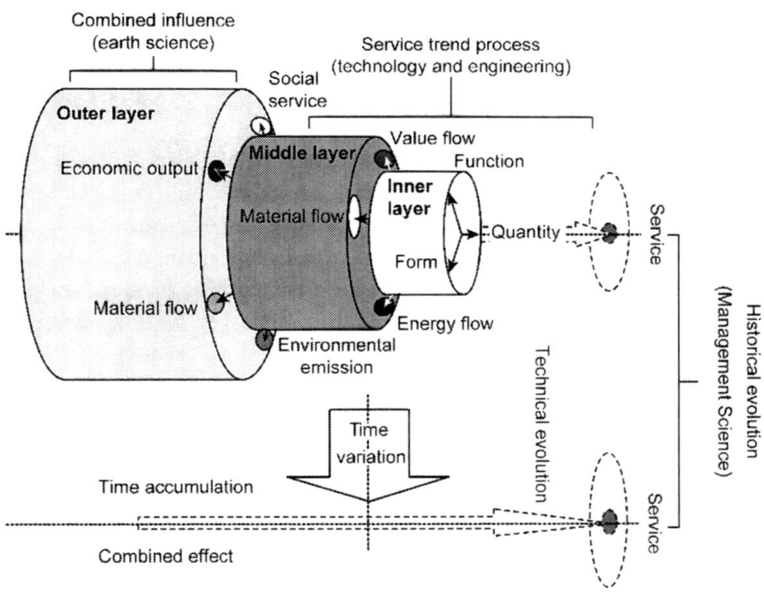

Fig. 10.3 Sketch map for the research framework of the service trend process of mineral resources. Reprinted from [27], with permission from Environmental Science & Technology (Chinese)

In terms of these processes and in static analysis, human service is regarded as the target point of the mineral resource service trend, the technical trend process from mineral resources to the implementation of final services is regarded as the technique-dimension axis, and, specific to a particular section, it is further divided into inner, middle, and outer layers. With the internal change of a substance as the core, the physical form, function, and its amount distribution relationship are focused on the inner layer. A basic production condition required by the implementation of the above material changes is valued in the middle layer, as well as the material, energy and value flows and the coupling relationships between them. In the outer layer, focus must be on the type and level of influence the process has on the external and on the external condition and inevitable outcomes of this process. During dynamic analysis, the demands' time-varying behavior must be fully considered, and based on the time-dimension axis, the trend process, and the historical variation process of their combined influences, the results are estimated to explore the internal action among elements within the technological sphere in this process and the external influence. Several typical human services are selected for case analysis and empirical study, and combined with the mineral resource utilization of a specific area, product design, service pattern, and technical and historical changes, the change rule of the service trend process of mineral resources is proved, and management and improvement suggestions are proposed.

(II) Example 2. Implementation guarantee for an urban circular economy during planning

Management is an important method to guarantee a circular economy, and planning is critical to execute integrated management for the to-be-implemented project. Planning the external factors that guarantee the implementation of a circular economy will be analyzed in this section.

As mentioned in Chap. 9, ecological city construction is one of the important factors in implementing a circular economy. Ecological city construction begins with planning. Here, the implementation guarantee in the planning of a specific ecological city completed by the School of the Environment of Beijing Normal University is used as an example to show the management factors considered. In this project, there are total of six safeguarding measures for the ecological city construction.

1. Establish a lead group to clarify the objectives and intensify organization and coordination

An ecological city is a trans-regional, trans-department, and multiindustry system; therefore, we must establish a special lead group where the group leader is the mayor and the members are the directors of the environmental resource community, urban construction community, development and reform commission, industry office, and other relevant departments. We must establish the objectives of the ecological city construction, achieve a unified arrangement with compound decisions, labor division, and cooperation among planning departments, and sign liability statements to form a promotion mechanism with level-to-level administration, departmental coordination, superior–subordinate relationships, and benign interaction.

2. Improve policies, perfect laws, and regulations

During ecological city construction, the economic model of a traditional economy (of which economic development has a significant part) is changed; it must intensify and reverse the concept of economic development by means of laws and policies. The country must establish economic policy guided by protection of the ecological environment to form a policy framework beneficial for plan implementation. For economic support, policy encouragement, and tax regulation, the social productive force factors must be guided to be beneficial to the ecological city construction, to guarantee and support the smooth implementation of the plan. Regions must establish and improve local regulations and systems related to environmental protection and ecological construction, stipulate methods, executive routines, and supervisory mechanisms for the ecological city construction, and clarify the post, right, responsibility, reward, and punishment measures of the institutional framework and related personnel to ensure the authority, importance, and executable power of the plan implementation.

3. Innovate management mechanisms and promote government service ability

Allow the government's macro-control ability in ecological construction to be fully employed, and on the basis of using the market resource's basic configuration, strengthen the prescriptive domination guidance function and synthesis coordination ability of the government, to practically solve the problems caused by the intersection of local protectionism and government function. Aggressively explore cross-regional governance models to develop a trans-regional coordination mechanism with information, mutual communication, unified action, and united supervision. Establish a green national economic accounting system to verify the resource environment and develop a related statistical index capable of fully reflecting the resource and environmental costs in economic development; furthermore, the objective of ecological environmental protection is integrated into the national economy of the government, an annual plan and medium- and long-term planning of social development.

4. Expand financing sources and reinforce funding capacity

The construction of an ecological city cannot be separated from its funding. All levels of government should strengthen their capital investment in ecological city construction, constantly expand its ratio in the total budget, and support land, water and other resources protection, ecological environmental management, and recovery activities. Establish multiple investment and financing mechanisms, and encourage and support the input of social funds into ecological city construction. Actively support bank loan applications for ecological projects, create equipment leasing and state special funds, issue enterprise bonds, and seek financing by listing the project on the stock market. Encourage different economic sectors and investors to support ecological city construction via various forms, including sole proprietorship, joint venture, contracting, lease, joint-stock cooperative system, and BOT. By employing these methods, sufficient funds for ecological city construction are guaranteed.

5. Strengthen information support, publicity, and education, and promote the management level of public participation

Effective and complete information is the basis of scientific management, and publicity and education provide the foundation to improve public awareness and practice. To support ecological city construction, a computer Internet platform should be applied to integrate the data information from each field relevant to the operation of an ecological city and to encourage the comprehensive integration of information resource sharing and data monitoring. An information disclosure and release system should be implemented to allow government, enterprise, and the public to understand the progress of city construction conveniently and efficiently. Multiple modes of propaganda and education at each level, such as training, radio, television, and public media, are fully applied to constantly promote understanding at all levels of leadership and the public concerning the circular economy, ecological environmental protection, and other aspects. Based on this, allow the leader to make scientific and reasonable decisions and integrate the construction progress of the ecological city.

6. Strengthen the team to promote technological innovative ability

Talent is the key to success in ecological city construction. Focusing on the key field and core technology of ecological city construction, accelerate talent cultivation and concurrently attract overseas talent to positively establish long-term technical cooperative relationships with institutions of higher learning and scientific research institutions at home and abroad to build the team for an ecological city construction. Encourage technological innovation, establish special funds, and support the development and application of new techniques and products. For cleaner production, ecological environmental protection, integrated use of natural resources, waste resource recovery, and ecological industry, positively develop, introduce, and promote domestic and overseas advanced technological achievements, execute the policy of developing a city by science and education, and prepare high technology-related preferential policies.

The guarantee system of ecological city construction embodies different types of elements, such as talent, management mechanisms, policy and other soft environment elements, information and production techniques and other hardware condition elements.

(III) Example 3. Water recycling system

Water resources are in great shortage in most regions of northern China. The recycling of water resources within specific areas becomes important for a circular economy. Here, the study of a specific scientific support project is used as an example to present the possible water recycling models for a city.

The case city is a specific coastal city; its water resource is formed from surface water, groundwater, seawater, and rainwater. Its water consumption system includes various centers of primary water production, supply from the source to the end users, and wastewater regeneration. In addition to supporting various functional zones (e.g., living quarters and industrial parks), we will further consider the relationships among agricultural systems, natural water systems of a city within an urban region and other areas. On the basis of coordinating the relationships among agricultural, domestic, industrial, and ecological water utilization, additional focus must be placed on the dirty water discharged from each functional zone. Further treatment techniques for wastewater are utilized to render the wastewater harmless and to discharge it into the environmental system. The available drainage water discharged from each functional zone is collected, the reclaimed municipal wastewater pipe network is established, and the water is transported to a functional zone with lower water quality requirements. Eco-technology is developed for domestic water, which supplements agricultural, ecological, and landscape water to form a compound network system of a stair-type water supply with multistage water utilization. This system is shown in Fig. 10.4.

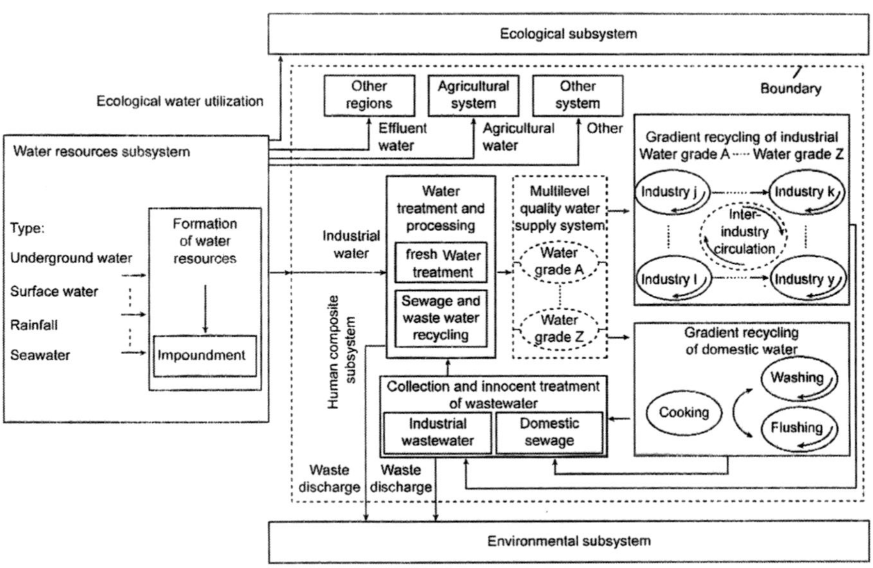

Fig. 10.4 Model framework of multistage and gradient water resource recycling based on a water quality gradient

It can be seen that to achieve efficient utilization of water resources in a city, we must promote the planning of urban water resources. Based on a specific water consumption system, a water recycling system is planned, and specific to each water consumption system, and based on the water quality requirements of different types of water consumption, the gradient utilization system of water resources among different water-consuming systems is planned. We must also prepare the corresponding hardware equipment (e.g., wastewater regeneration equipment) and infrastructure (e.g., reclaimed water pipeline). However, we must perform scientific appraisal and practice of the new technology research and development, design, system operation, and city management. All these become necessities to guarantee the highly efficient operation of an urban water system.

10.2.3 What Should Be the Structure of a Circular Economy Guarantee System?

(I) Structure of a circular economy guarantee system

As previously mentioned, a circular economy involves various socio-economic fields and has a complex multidimensional structure. For example, regarding system types, the material processing and transformation process primarily occur in an industrial system, while after the product is developed with specific service functions, its use occurs in a social life system. For material source and final trend, the natural resources and metabolic waste are finally discharged into the environmental system; therefore, they are related to resource possession, distribution and environment capacity and subordinate to the resource-environmental system. Technique includes the process from the natural system material to the product that satisfies the specific needs of human beings. Technique not only specifies the method to change a material's property via research and development but must guarantee the realization of the material's change in physical form and function based on the specifications of the design. Finally, the actual production process is required to complete these processes. In general, it will be related to many factors, as shown in Fig. 10.5. The majority of the dimensional variation relations have been discussed in previous chapters, such as the selection of material type and the life cycle phase of the material. Another portion of the variation relation is embodied in several previous examples, such as the multidisciplinary problem showing the resources trend and the multilevel administrative measure problem embodied in urban planning practice.

(II) Complexity of a circular economy guarantee system

From Fig. 10.5, it can be observed that the circular economy guarantee system can be composed of various subsystems from different fields of study. It has a very complex structure; its composition not only varies, but different fields are associated internal with the system, which influences the functions of the entire guarantee

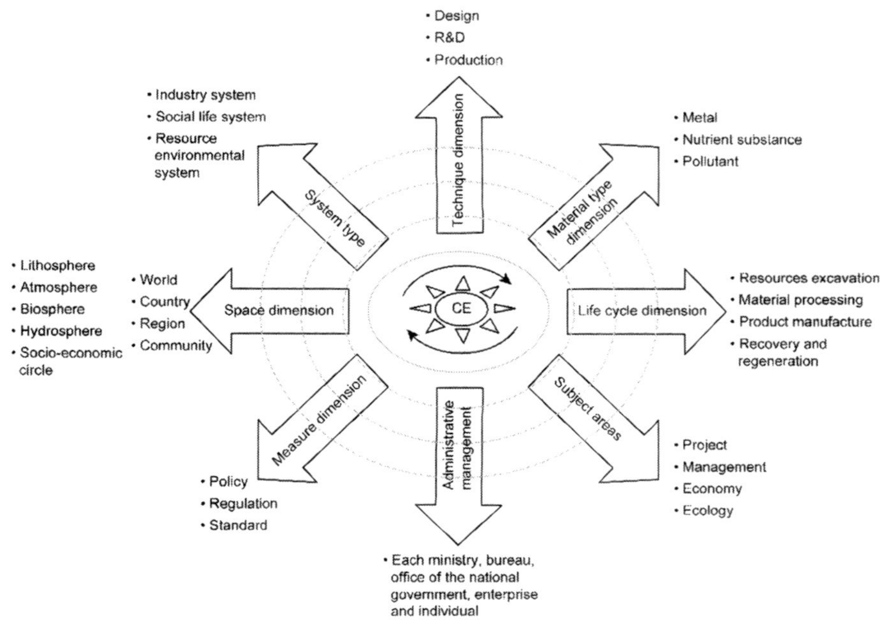

Fig. 10.5 Complexity of the relationship between system structure and elements of the circular economy

system. Additionally, the structure of the entire guarantee system varies based on social and economic development, which will have different effects during different stages of the circular economy; therefore, the entire circular economy guarantee system will also have different effects. For example, in the initial stage of a circular economy, improvement of the circulation rate might be the key emphasis and a management, technical, or waste recovery policy would be prepared to promote material recycling. In the medium stage of a circular economy, the key emphasis might be improvement of the consumption model, reduction of the economic growth rate, and stabilization of development quality. In addition, as there are very different natural resource endowments and environmental conditions in different regions, the developmental level of their human society economy is also different, which makes the development situation of a circular economy and its guarantee system distinctive; its future development is unpredictable. Thus, the circular economy guarantee system shows significant complexity because it not only includes a variety of factors in the natural system but also includes many factors in the human society economy. There are complex structures, diversified functions, different interactions among elements and spatiotemporal dynamic evolution, which reflect the interactive relationship between human beings and the environment.

Because the circular economy guarantee system has complexity, its evolution rules must be disclosed by complexity science. For the convenience of understanding the future tendency of the circular economy guarantee system, complexity science is briefly described below.

Complexity science is a branch of science formed in the 1980s to address the complex system problem. It is a new interdisciplinary branch of science; its research object is a complex system, its primary task is to reveal and explain the operation of a complex system, and its primary purpose is to improve people's ability to understand, explore, and change the world. Complexity science has the following essential features: ① to deeply understand the nonlinear interaction relationships of system complexity and to believe that the nonlinear interaction relationships among elements constituting the system are the basic cause of the system's infinite variety, unpredictability, and significant difference; ② to fully recognize the uncertainty of the system's operational result and to believe that there exist certain and uncertain relationships among the system components and system function during operations; the certainty and uncertainty are mutually associated and transformed to form the colorful real world; ③ to believe that the system has self-organization, i.e., the system can achieve self-creativity, self-evolution, independently change from chaos to order, and can form or possess a specific intrinsic-ordered structure; system opening, far from equilibrium, nonlinear interaction, and fluctuation are the basic conditions of the formation of self-organization; "Survival of the fittest" is an example of a self-organization process; specifically, an organism in the natural world achieves evolutionary development among different species through autogenous regulation of the ecological system; and ④ to fully recognize the properties, characteristics, behaviors, and functions that are owned by the entire system but not by each component, i.e., emergent property. Put differently, each component constituting the system has its own property, characteristic, behavior, and function, and when these components are integrated into an entire system, it will result in different properties, characteristics, behaviors, and functions. Complexity sciences will assist us in understanding the world more deeply and to obtain a more effective method to solve practical problems. Its appearance indicates the development of system science at a new level. Additionally, we believe that it has an important effect on the improvement of a circular economy guarantee system.

10.3 Method to Guarantee the Implementation of a Circular Economy

The guarantee of a smooth implementation of a circular economy is safeguarded in different countries via multiple management methods, such as law, policy, and regulation. The legal safeguard is the highest of numerous safeguarding measures. Here, on the basis of understanding the development status of the Japanese circular economy, Japan and our country are used as examples to specify several core provisions of the circular economy promotion law in China.

10.3.1 Overseas Example: Development Status of the Japanese Circular Economy

(l) Background of the Japanese circular economy

In Japan, the circular economy was initially implemented to solve the problem of progressively increasing wastes. In the 1970s, with the economic development and population increase in Japan, the discharge amount of household refuse increased constantly, resulting in increasing saturation of the refuse landfill; therefore, Japan had to take measures to solve their garbage problem. Concurrently, a significant amount of the wastes provides a material basis for their input as secondary resources. To promote the reduction of garbage and waste recycling, in 1991, the *Promotion Act for the Effective Utilization of Resources* was issued in Japan to provide a legal requirement for the recycling of waste resources.

With the constant implementation of a circular economy and specific to the types of renewable materials, e.g., automobile, food, and building material, special laws for a circular economy were successively issued in Japan, as shown in Table 10.1. A circular economy is brought into law to effectively reduce the amount of garbage and improve the waste recycling rate.

For example, after issuing the *Recycling Method of Household Appliances* in 1998, household appliance manufacturers assumed the obligation to reclaim and utilize waste household appliances. Television sets, refrigerators, air conditioners, washing machines, and other large household appliances are compulsively recycled, with recovery rates of 55, 50, 60, and 50%, respectively. It is further stipulated

Table 10.1 List of Japanese laws related to a circular economy

Year	Name of law	The level of legal system
1970	Solid Waste Management and Public Clean Act	Comprehensive law
1991	Promotion Act for the Effective Utilization of Resources	Comprehensive law
1993	Basic Environment Act	Basic law
1995	Method for Promoting Container and Packaging Classification Collection	Special law
1998	Recovery Act for Household Appliances	Special law
2000	Basic Law for Promoting the Formation of a Recycling-oriented Society	Basic law
	Food Recycling Method	Special law
	Green Procurement Law	Special law
	Building Materials Recovery Law	Special law
2001	Extraordinary Measures Law for the Proper Treatment of Polychlorinated Biphenyl Wastes	Special law
2002	Automobile Recycling Law	Special law

by law that consumers shall assume partial costs of the regeneration treatment; concurrently, enterprises are encouraged to actively participate in the recovery and reuse of waste.

(II) Implementation measures of the Japanese circular economy

The implementation measures of the *Basic Law of Promoting the Formation of a Recycling-oriented Society* in Japan are used as an example.

Based on the *Basic Law of Promoting the Formation of a Recycling-oriented Society* combined with the *Johannesburg Plan of Implementation* formed at the Johannesburg world summit conference for sustainable development in September 2002, to accelerate the transformation into a sustainable production and consumption pattern, the *Basic Plan to Boost the Formation of a Recycling-oriented Society* was prepared by the Japanese government in March 2003. The plan proposes numerical indices and a series of strategic measures for the implementation of a recycling-oriented society. The following three key indicators were to be achieved by the circular economy by 2010: Resource productivity (resource productivity equals output value divided by the input of natural resources) should reach 390,000 Yen/t, improving by 40% relative to 2000; the cyclic utilization rate should reach 14%, improving by 40% relative to 2000; and the final disposal amount of garbage should be 2.8 million tons, equivalent to 50% of the amount in 2000. In addition, compared with 2000, the daily amount of household refuse per person reduced by approximately 20%, and the final treatment capacity of industrial garbage reduced by approximately 75% relative to 1990. The market size and employment within the recycling-oriented society were twice that of 1997.

The promotion of a Japanese recycling-oriented society depends not only on legislation but also on the positive participation of all walks of life. A country must actively prepare national policy, an environmental audit system, and a reward and punishment system. An enterprise must dispose of wastes in strict accordance with the legal provisions. It was not only attempted to develop cleaner production techniques, but actively applied the international ISO 14001 environment management system certification. Through 2000, approximately 3500 enterprises have obtained this certification and each enterprise prepares an environmental impact report to disclose its environmental condition to the public. Each industry must actively supplement the social venous industry to form the material's cyclical pattern of "natural resources-product-renewable resources"; each region must build material metabolism and symbiosis among enterprises to form an eco-industry park and achieve zero release of pollutants; and each society must actively advocate the construction of a recycling-oriented society, assist the public in changing their lifestyle by means of community organizations, media publicity, and supervision, and coordinate the disposal of wastes to allow the ideas of a circular economy to obtain popular support.

(III) Basic characteristics of the Japanese circular economy

The implementation of the Japanese circular economy has the following basic characteristics:

First, the Japanese circular economy is implemented at different levels and has a different emphasis in each area. Enterprises must focus on developing cleaner production techniques, solving environmental pollution, achieving energy replacement, and performing activities in accordance with the international ISO 14001 environmental management system certification. Regions must focus on building industrial metabolism and symbiosis among enterprises to form an eco-industry park. Industries must focus on developing a social venous industry to form the material circulation mode of "natural resource-product-renewable resource." The entire society must support the objectives of building a recycling-oriented society and limiting the consumption of natural resources to achieve a minimum environmental load. In addition, the Japanese government further attempts to achieve the complementarities of relevant policies prepared by each department and to promote the close cooperation of these departments via cabinet council meetings, ministerial conferences, and department liaison meetings.

Second, the core of the Japanese circular economy is waste recycling. As a developed country, the national high income results in high consumption and causes high waste discharge. Waste recycling can not only solve the problems of waste disposal site and environmental pollution, but can ease the situation of an "island country" resource shortage. It becomes the core of developing the circular economy in Japan. The existing eco-industry parks in Japan are primarily engaged in waste recycling. The wastes involved in materials recovery and recycling include general waste and industrial waste, e.g., automobiles, household appliances, printers, ink boxes, beverage containers, and papers.

Third, the development of the Japanese circular economy is based on technical progress. Since the circular economy has been implemented, the Japanese government has allocated a significant amount of funds to perform technical research of waste recycling. It is reported that the environmental protection-related budget allocated by the Japanese Diet exceeds 1 trillion Yen every year, including a budget of 150 billion Yen for waste disposal and recycling. In 2003, the total provincial budget used for implementing the recycling-oriented society was 445.233 billion Yen. In 2004, the METI's financial budget was 2.47 billion Yen; 1.54 billion Yen was used for the construction of an eco-industry park, 880 million Yen was used to promote the recycling of automobiles, household appliances, and containers, and 550 million Yen was used for industrial development and social capital construction of durable maintenance system. Additionally, the government provides financial support to communal facilities of the recycling-oriented society. For example, through close cooperation among enterprises, administrative departments, and universities, the validation research area of the Kitakyushu eco-industry park will be integrated with an organization that performs research on the development of top waste disposal techniques and recycling technology to be the development base of relevant environmental technology.

10.3.2 Our Country's Circular Economy Law

(1) Development of China's circular economy

In our country, the circular economy experienced the following stages:

At the end of the 1980s, China actively participated in the implementation of the *Cleaner Production Plan* prepared by UNEP. The prevention and control of industrial pollution began the strategic transformation from end treatment to cleaner production. Additionally, the *Cleaner Production Promotion Act* prepared in our country was the first among developing countries.

At the end of the 1990s, our country introduced the new concept of a circular economy from abroad, and it quickly gained the attention of the supreme leadership. At the global environment fund member countries conference in 2002, Comrade Jiang Zemin delivered an important address that stated: "only by this method of a circular economy based on the most effective utilization of resources and environmental protection, can we achieve sustainable development." In March 2003, Comrade Hu Jintao stated that "we should accelerate the transformation of economic growth to realize the development of a circular economy through regional economic development and product production, so that resources can be utilized most effectively. We should maximize the reduction of waste discharges to allow zoology to move into a virtuous cycle, and we should attempt to build an environmental protection model city, ecological demonstration zone and ecological province." This policy received responses from various regions, and the implementation of a circular economy became very popular throughout the country. On June 29, 2002, the *Cleaner Production Promotion Act of the People's Republic of China* was approved with revision during the 28th meeting of the 9th NPC Standing Committee.

In July 2005, the State Council officially released *Several Suggestions for Accelerating the Development of a Circular Economy by the State Council* to propose guidelines for the purpose, key field, and management policy of circular economy development in China. The governments at all levels and relevant departments were requested to regard the development of a circular economy as the important guiding principle to prepare related planning, and the idea of a circular economy was used to guide the preparation of the "11th Five-Year Plan," various regional planning, urban master planning, sustainable utilization of mineral resources, energy conservation, water saving, integrated use of natural resources, and other special planning. The promotion plans of the circular economy are prepared and implemented. The circular economy becomes the major development strategy of the country. In October of the same year, the National Development and Reform Commission, State Environmental Protection Administration, and six other departments jointly selected 42 enterprises in seven major industries (e.g., steel, nonferrous metal, and chemical engineering), 17 units in four major fields (e.g., renewable resources recycling), and 13 different types of industrial parks to execute the first pilot projects of a circular economy in our country, to explore the development patterns of a circular economy and to promote the establishment of resource recycling mechanisms.

In September 2008, the *Circular Economy Promotion Act of the People's Republic of China* was officially issued, indicating that China's circular economy activities were moving forward as planned. In 2012, the *Cleaner Production Promotion Act of the People's Republic of China* was revised. To date, four sets of pilot projects of a circular economy have been implemented in our country, indicating that the circular economy is increasingly mature.

(II) What are the contents stipulated by China's circular economy law?

A circular economy management system, material reduction, recycling, incentive measures, legal liability, and other basic components are specifically stipulated by our country's circular economy promotion act, and the purpose is to effectively promote the development of the circular economy, improve resource use efficiency, protect and improve the environment, and achieve sustainable development. These elements are further discussed in the following sections.

1. Management system

The basic management system for the implementation of the circular economy in our country is explicitly stipulated in the *Circular Economy Promotion Act of the People's Republic of China*. The contents include: ① Development planning system—The comprehensive management departments for circular economy development at all levels are requested to jointly plan the development of the circular economy in local administrative regions along with the environmental protection department and report to the local government for release and implementation after approval. The development planning of a circular economy should include planning targets, range of application, main content, key tasks, and safeguarding measures, and must stipulate the resource output capacity, refuse reclamation, and recycling rate. ② Preparation of total quantity control—The local people's government above the county level is requested to plan and adjust the industrial structure within the local administrative region to promote the development of a circular economy according to the control standards of the principal pollutant discharge, construction land, and total amount of water consumption within the local administrative region, as issued by the superior people's government. A newly built, rebuilt, and expanded construction project must be in accordance with the requirements of the control standards of the principal pollutant discharge, construction land, and total amount of water consumption within the local administrative region. ③ Preparation of an indicator assessment system—The comprehensive management department for circular economy development of the State Council is requested to establish and improve the indicator assessment system of the circular economy jointly with statistical, environmental protection, and other departments within the State Council. According to the main assessment criteria of a circular economy, the superior people's government should regularly assess the development status of the circular economy and should regard the performance as the content of the examination and evaluation to the local people's government and its principal. ④ Waste recovery system—The manufacturing enterprise whose product or packaging is listed in the compulsory recycling directory must recycle the abandoned products or packaging. If available, the recycled products are utilized by the manufacturing enterprise; if unavailable due to lack of technical–economical conditions, the manufacturing enterprise is responsible for harmless disposal of the product or packaging. If the seller or other organization is entrusted by the manufacturing enterprise to recycle or a waste recovery or disposal enterprise is entrusted to recycle, the commissioned party should be responsible for recovery, utilization, or disposal in accordance with relevant laws, regulations, and agreements of contract. Additionally, the consumer of this type of product and its packaging is requested to surrender the discarded elements to the producer, entrusted seller, or other organization for recovery. ⑤ Supervision and management system—The country is requested to supervise and control the energy and water consumption of the key enterprises whose annual integrated energy and water consumption exceed the total quantity stipulated by the state for their industries (i.e., steel, nonferrous metals, coal, electric power, and petroleum processing). The energy consumption is executed in accordance with the Energy Conservation Law of the People's Republic of China. ⑥ Statistical system—The country is requested to establish and improve the statistical system of the circular economy to strengthen the statistical management of resource consumption, comprehensive utilization, and waste production. The major statistical index should be regularly released to the public.

2. Minimization

China is faced with a severe shortage of resources. The per capita quantity of 45 major mineral resources is less than one half of the world average, and the per capita reserve of petroleum, natural gas, bauxite, and other important mineral resources is less than 10% of the world average. To conserve resources and reduce waste discharge, it is expressly stipulated in the *Circular Economy Promotion Act of the People's Republic of China* that under the premise of technical feasibility and economic rationality, the development of a circular economy should obey the principle of "minimization priority."

Minimization refers to the reduction of resource consumption and waste production during production, circulation, consumption, and other processes. To achieve "minimization," the major measures are as follows: ① Regularly release the list of the encouraged, limited, and eliminated techniques, processes, equipment, materials, and products. For example, for flue gas desulfurization in a large-scale coal-fired industrial furnace, the use of an inner circulation multistage spouting fluidized flue gas desulfurization technique is encouraged in the *List of the Environment Protection Technology Encouraged by the State in 2008*. ② Industrial water saving, seawater utilization, reclaimed water reuse, and tap water saving—For example, industrial water saving is expressly stipulated in our country's circular economy promotion act; first, for industrial enterprise water conservation and measuring control, the industrial enterprise is requested to utilize advanced or suitable water-saving technology, technology and equipment according to the relevant state water-saving product catalog and technological policy. Second, prepare a water-saving plan for the enterprise according to the water quota stipulated by the state or relevant water-saving index. Third, the enterprise is requested to strengthen water-saving management to practically implement the water-saving plans and measures via establishing and perfecting the water-saving management system and management system of the enterprise. ③ For mineral resource conservation and protection, first, consider resources, technology, economy, security, the environment and other factors when mining mineral resources, perform overall planning, prepare a reasonable development and utilization program, and utilize rational mining sequences, methods, and mineral processing technology. Second, the mining license issuing authority is requested to review the indicators, such as mining recovery, mining dilution rate, ore dressing recovery percentage, water circulation utilization of the mine, and land reclamation rate in the development and utilization program submitted by the applicant. If an applicant is not qualified, do not issue a mining license. Finally, the mining license issuing authority is requested to reinforce the supervision of mineral resource mining according to laws, to change the previous situation of valuing review but relaxing supervision. For paragenic and associated ore, it is further stipulated in this law that when a mining enterprise possesses practical technical capability and economic feasibility, it should perform comprehensive mining and rational utilization of the industrial value-possessing paragenic and associated ore while mining their primary minerals. When a mining enterprise temporarily does not possess the practical technical capability or economic feasibility, safeguarding procedures must be applied to the mineral products that must be mined simultaneously but not utilized immediately, as the tailings contain useful constituent, thus preventing resource loss and ecological damage.

3. Recycle and reclamation

"Recycle" refers to the waste that is directly used as a product or is regarded as a product for use after being repaired, renovated, or reproduced, or the entire or portion of the product is regarded as a component of other products for use. "Reclamation" refers to waste that is directly used as raw material for use or recycle. Waste recycling and reclamation can greatly conserve resources and reduce waste discharge; it is an important component of a circular economy. Waste recycling and reclamation include a regional circular economy, comprehensive utilization of industrial waste, comprehensive utilization of waste heat/pressure, and construction of a waste recovery system.

The regional circular economy is related to the individual enterprise and is used to solve the problems that cannot be solved by the enterprise itself at a higher system level, for example the industrial distribution or allocation problems of the production factors among different industries. Article 29 of the *Circular Economy Promotion Act of the People's Republic of China* stipulates "the people's government above the county level should develop overall planning for the regional economic layout and rationally adjust the industrial structure to promote enterprise cooperation in the fields of integrated use of natural resources and to achieve efficient utilization and recycling of resources. Various types of industrial parks should organize the enterprises within their own area for the integrated use of natural resources, to promote the development of a circular economy. The country should encourage the enterprises in each industrial park to perform waste exchanges, energy step utilization, intensive land use, classification and recycling of water and to jointly use infrastructure and other related facilities. Environmental impact assessment should be performed on the newly built and renovated industrial parks according to laws, and ecological protection and pollution control measures should be employed to ensure that the environment quality within the area reaches the specified standards". Therefore, in the development of a regional circular economy, the people's government above the county level should perform overall planning for the entire regional circular economy and should demand each enterprise participate in efficient coordination. Based on the principle of circular economy development, each industrial park must actively promote the comprehensive utilization of the enterprise in the aspects of resource, energy sources, land, water, and infrastructure and protect the local ecological environment during the production process. By guaranteeing the smooth implementation of the regional circular economy, regional resource saving and environmental awareness, economic growth and the harmonious development among population resources and the environment can be achieved.

Industrial waste refers to waste produced during industrial production activity. Due to the diversity of industrial sectors, there are a great variety of industrial wastes in large quantity. To strengthen the comprehensive utilization of industrial waste to conserve resources and reduce waste discharge, it is important to develop a circular economy. It is stipulated in the *Circular Economy Promotion Act of the People's Republic of China* that "the enterprise should perform comprehensive utilization of the industrial wastes produced during the manufacturing process, such as, fly ash, coal gangue, tailings, barren rock, scrap and exhaust gas in accordance with national regulations." According to the regulation, the enterprise is requested to utilize industrial solid wastes based on economic and technical conditions and to perform classified storage or harmless treatment of the temporarily unavailable or unavailable wastes.

Waste heat/pressure refers to heat and pressure that has not been completely utilized after being produced in a boiler, kiln, and converter during industrial processes. Waste heat/pressure is a type of energy itself and is a normal phenomenon that occurs during industrial processes. If it cannot be fully utilized, it would cause an enormous energy waste. Article 32 of the *Circular Economy Promotion Act of the People's Republic of China* stipulates "the enterprise should employ advanced or suitable recycling technology, processes and equipment to perform comprehensive utilization of the waste heat/pressure produced during the production process. The grid-connected power generation project of the construction waste heat/pressure, coal bed methane, coal gangue, coal slime, garbage and other fuel with low heating value should be provided with administrative licensing or submitted for filing in accordance with the laws and regulations of the State Council. According to national regulations, the power grid enterprises should sign the grid-connected agreement with the enterprise that generates electricity via the comprehensive utilization of resources to provide surfing service and purchase the full amount of on-grid energy of the grid-connected power generation project." It can be observed that the regulation not only demands the comprehensive utilization of waste heat/pressure for the manufacturing enterprise but also requests power grid enterprises to receive and purchase the manufacturing enterprise's electric power generated via waste heat/pressure.

Waste recovery is an important method to improve the resource utilization rate, protect the environment, and develop a circular economy. Article 37 of the *Circular Economy Promotion Act of the People's Republic of China* stipulates "the construction of a waste recovery system is encouraged and promoted by the country. Based on urban and rural planning, the local people's government should rationally arrange the waste recovery website and trading market and should support the waste recovery enterprises and other organizations to perform waste collection, storage, transportation and exchange of information. The trading market waste recovery should be in accordance with the national environmental, safety and fire protection regulations." It can be observed that the regulation clarifies the management function of the country and region in the construction of waste recovery systems and also clarifies the requirements of the waste recovery trading market.

With the rapid development of the electronic information industry, the waste electrical and electronic products and other specific products have increased accordingly. Article 38 of the *Circular Economy Promotion Act of the People's Republic of China* stipulates "the dismantling or recycling of waste electrical and electronic products, scrapped motor vehicle/ship, scrap tire, waste lead-acid battery and other specific products should be in accordance with the relevant laws and administrative regulations." Article 39 stipulates "when the reclaimed electrical and electronic products are sold after being repaired, the recycling product standards and labeling of recycled products in a significant position must be followed. If the reclaimed electrical and electronic products must be dismantled and recycled, they should be sold to a qualified dismantling enterprise." It can be observed that the regulation clarifies the recycling scope of the waste electrical and electronic products, and the enterprise's qualification requirements in this process. In our country, the associated laws further include *Prevention and Control Act for the Environment Pollution of Solid Wastes, Prevention Management Method for the Environment Pollution of Electronic Wastes, Regulations on Scraped Car Recovery, Control Regulations on Preventing Environment Pollution of Ship Scrapping,* and *Recycling Control Regulations on Waste Electrical & Electronic Products.*

4. Incentive measures

To promote the sound development of a circular economy, the incentive measures to be used are clearly stipulated in the *Circular Economy Promotion Act of the*

People's Republic of China and primarily include: ① The establishment of special funds for a circular economy to provide financial support for the related technical innovations—Article 42 of the *Circular Economy Promotion Act of the People's Republic of China* stipulates "the special relevant funds for the development of a circular economy are established by the State Council, province, autonomous region and municipality people's government to support the science-technology R&D of the circular economy, demonstrate and popularize the circular economy technology and product and implement a major circular economy project and information service for the development of circular economy." The specific methods are prepared by the financial sector along with the integrated management division for the development of a circular economy of the State Council and other related departments. The regulation specifies the implementation subject of the financial support and the supporting content. ② Clarify the major field of investment and provide financial support and tax preference—Article 45 of the *Circular Economy Promotion Act of the People's Republic of China* stipulates "when the investment plan is prepared and implemented by the comprehensive management department for the development of economic development of the people's government above the county level, energy conservation, water saving, land saving, material saving, comprehensive utilization of resources and other projects must be identified as key fields of investment. For energy conservation, water saving, land saving, material saving, comprehensive utilization of resources and other projects conforming to national industrial policies, financing institution should offer senior loans and other credit aids, and should actively provide supporting financial services. For production, import, sales enterprises or those using the technology, process, equipment, material or product that is listed in the elimination directory, financing institution shall not provide any form of credit support." For example, in 2006, the circular economy and resource saving demonstration projects were awarded the key investment support of the central government. ③ Recognition awards—Article 48 of the *Circular Economy Promotion Act of the People's Republic of China* stipulates "the people's government above the county level and the related department should give commendations and awards to the entities and individuals making remarkable achievements in the fields of circular economy management, science and technology research, product development, demonstration and extension. The enterprises and institutions should give commendations and awards to the entities and individuals making outstanding contributions in the development of a circular economy." For example, to promote the establishment of a circular economy industry system, more than 1 million Yuan is paid by a specific bureau of finance to reward the enterprise and major project that makes outstanding contributions to the circular economy, energy conservation, and emissions reduction every year.

5. Legal liability

Legal liability is the compulsory binding means for the promotion of circular economy development. The supervision management department of a circular economy and the design, circulation and using departments of the product, technology, process, equipment, and material are explicitly stipulated in the *Circular Economy Promotion Act of the People's Republic of China*. ①For the supervision management department, Article 49 of the *Circular Economy Promotion Act of the People's Republic of China* stipulates "if there is an act in violation of this law in the comprehensive management department for the development of a circular economy of the people's government above the county level or other related department, or an unlawful act is not investigated and treated thereby after receiving warnings, or there exists another act of not performing supervision duty in accordance with the law, the related department of the local people's government or the superior people's government and the direct governor and other responsible personnel should be punished according to the laws." ②For the production, circulation, and using departments, Article 50 of the *Circular Economy Promotion Act of the People's Republic of China* stipulates "for the product or equipment whose production and sales are listed in the elimination list, it should be punished in accordance with the regulations of the *Product Quality Law of the People's Republic of China*. Those using the eliminated technology, process, equipment and material should be instructed to stop by the comprehensive management department for the development of a circular economy of the people's government above the county level, the illegal equipment and material should be confiscated, and 50,000–200,000 Yuan should be fined. For a case of gross violation, the comprehensive management department for the development of a circular economy of the people's government above the county level should provide advice and submit direction to the local people's government to instruct cessation or closure of the business in accordance with the provisions of the State Council. If the equipment, material or product listed in the elimination directory is imported in violation of the law, Customs should be instructed to return the cargo and the enterprise should be fined 100,000–1 million Yuan. For an unclear importer, the carrier should bear the responsibility of the returned cargo or assume the relevant treatment cost." ③The *Circular Economy Promotion Act of the People's Republic of China* stipulates the design department is not allowed to use the toxic and harmful substances listed in the country's prohibited directory; if the provision is violated, there will be an investigation, and legal liability will be attributed. "In violation of this law, for electrical and electronic and other products possibly causing environmental pollution during their dismantling or disposal process, the toxic and harmful substances listed in the country's prohibited directory are used, and the enterprise should be instructed to correct the violation within a time limit provided by the product quality supervisory authority of the local people's government above the county level. If it is not corrected within the time limit, 20,000–200,000 Yuan is the fine. For a serious violation, the relevant circumstances should be reported by the product quality supervisory authority of the local people's government above the county level to the local industrial and commercial administrative department, and the enterprise's business license should be revoked by the industrial and commercial administrative department."

10.3.3 Comparison Between China and Foreign Countries' Circular Economy Promotion Acts

The *Circular Economy Promotion Act* issued by China was prepared on the basis of objectively understanding the environmental resource questions in China and absorbing international advanced experience.

Comparing *China's Circular Economy Promotion Act* with Japanese and German circular ecology legislation, several identical points can be found. For instance, the legislation logic and mode of the Japanese *Basic Environment Act* and *Basic Law of the Establishment of a Recycling-oriented Society* are referenced in articles 8 through 11 of the *Circular Economy Promotion Act*, specifically, the provisions concerning the responsibilities of the country, enterprise, individual and industry. For the implementation of the entire process of a circular economy stipulated in the *Circular Economy Promotion Act*, the German producer liability system is referenced in the packaging minimization, reuse and recycling stipulated in Article 15.

In addition, there are many differences in the relevant laws between China's *Circular Economy Promotion Act* and Japan's act as follows: ① The management content of China's *Circular Economy Promotion Act* states that resources and the environment are similar; e.g., the utilization of waste resources is emphasized in the basic law of a circular economy in China, while resource protection and environmental protection are similar in the *Circular Economy Promotion Act*. This not only reflects general promotion but also reflects a major breakthrough in China's resource conservation. The general promotion is embodied in several aspects of the policy, objective, principle, and management system in "general provisions" and "basic management system." A major breakthrough is reflected in the "minimization" and "reuse and recycling" principle requirements proposed for product design, packaging, water saving, oil saving, mineral resources saving, building material saving, land saving, irrigation water saving, government saving, food saving, forestry saving, or resources reuse or recycling, which allows the legal construction of the circular economy to address the key activities and major aspects of circular economy development. ②China's *Circular Economy Promotion Act* has a superior and complete management system; e.g., by means of the point–line–plane three levels of responsibility, it is effectively unified, and the government's responsibility is highlighted. The responsibility at the point level refers to the business and individual duties specified in articles 9 and 10 of the *Circular Economy Promotion Act*, respectively. The responsibility at the line level refers to the duty of each industry specified in Article 11 of the *Circular Economy Promotion Act,* and the responsibility at the plane level refers to the duty of the local administrative region (e.g., development area) at each level of the government specified in Article 8 of the *Circular Economy Promotion Act*. The government's responsibility is described in Article 8 of the *Circular Economy Promotion Act*; in addition, the government's planning obligation and industrial restructuring duty are further stipulated in the "basic management system" of Chap. 2. Additionally, the superior government's examining and assessment system of the circular economy to the subordinate government is stipulated in Article 14 of the *Circular Economy Promotion Act*.③ It embodies the fundamental realities of our country from the aspect of resources and considers the actual mode of the environmental resources administrative management system and environmental resources management in China. For example, it greatly embodies the situation of rapid growth of our country and the terminal values of promoting economic growth, improving resource use efficiency and protecting and improving the environment. At a deep level, what is emphasized can not only meet consumer demands but can also control the consumption of resources and satisfy people's ever-increasing ecological and living demands, to finally build a sustainable society with a small environmental load.

Classroom Discussions and Assignments

Topic Discussion

The circular economy guarantee system is related to multiple disciplines, including natural and human sciences; therefore, it must provide effective guarantees from different perspectives to promote its smooth development based on the plan. A variety of unsatisfactory problems encountered during the implementation of a circular economy also shows that there are currently many shortcomings in the circular economy guarantee system. Consider and discuss the following topics:

Choose a specific safeguarding measure in our country's existing circular economy laws, and consider how it provides the guarantee in the circular economy.

Alternatively, combined with the typical events in the implementation of a familiar circular economy, provide disadvantages and propose improvement suggestions to the existing guarantee system.

Homework and Presentation in Class

Specific to the topics discussed in class, a scientific research group is the work unit for after-class study to develop a group report to share in class. As the topic is expansive, personal innovative ideas are considered in the grade. The grading scale is shown in Table 10.2.

Table 10.2 Grading of the group report in class

Report title	Process of topic selection (maximum 10 points)	Research process (maximum 10 points)	Presentation skill (maximum 10 points)	Innovation (maximum 10 points)	Total points

Recommended and Referenced Literatures

1. Ayres, R.U. 2001. *Industrial metabolism: Theory and policy, the greening of industrial ecosystems*, 23–37. Washington, DC: National Academy Press.
2. Geng, Y., J. Fu, J. Sarkis, et al. 2012. Towards a national circular economy indicator system in China: An evaluation and critical analysis. *Journal of Cleaner Production* 23 (1): 216–224.
3. Li, X.N. 2011. LI Y F. Driving forces on China's circular economy: From government's perspectives. *Energy Procedia* 5: 297–301.
4. Ma, S.H., Z.G. Wen, J.N. Chen, and Z.C. Wen. 2014. Mode of circular economy in China's iron and steel industry: A case study in Wu'an city. *Journal of Cleaner Production* 64 (1): 505–512.
5. Su, B.W., A. Heshmati, Y. Geng, and X. Yu. 2013. A review of the circular economy in China: Moving from rhetoric to implementation. *Journal of Cleaner Production* 42: 215–227.
6. Graedel, T.E., and B.R. Allenby. 2003. *Industrial ecology*, 2nd ed. Upper Saddle River, NJ: Prentice Hall.
7. Graedel T E, Allenby B R. Industrial Ecology. Translated by Shihan. Beijing: Tsinghua University Press, 2003.
8. Guan, X. 2008. *Tax policy study on promoting the development of circular economy*. Beijing: China Tax Publishing House.
9. Huang, X. 2006. *Research methodology on complexity sciences*. Chongqing: Chongqing University Press.
10. Li, C., Y. Liu, and G. Shao. 2016. Study on the security system of emergency rescue forces for dangerous chemicals. *Fire Science and Technology* 35 (2): 279–282.
11. Li, H., and Y. Ren. 2006. China's policy framework on developing circular economy. *Environmental Economy* 4: 48–51.
12. Yunyan, Li. 2008. *Operating mechanism of circular economy-market mechanism and government action*. Beijing: Science Press.
13. Wei, Liu. 2009. *Establishment of scientific development and circular economy mode*. Beijing: China Economic Publishing House.
14. Lu Zhong wu. 2010. *Foundation of industrial ecology*. Beijing: Science Press.
15. Dan, L.V. 2009. *Study on the promotion mechanism of government-oriented circular economy*. Beijing: China Social Sciences Publishing House.
16. Ren Y, Zhou G, etc. Development Mode and Policy of Chinas Circular Economy. Beijing: China Environmental Science Press, 2009.
17. Shang, H., and S. Zhou. 2008. *Evaluation and policy design of circular economy development*. Beijing: China Finance Economics Press.
18. Sun, C., and J. Zhou. 2007. *Institutional innovation and circular economy*. Beijing: Economic Daily Press.
19. Sun, Y., and J. Zhao. 2008. *Interpretation of the circular economy promotion act of the People's Republic of China*. Beijing: China Legal Publishing House.
20. Tao, C. 2010. *Road of Green-Excellent case silhouette of ningbo circular economy*. Hangzhou: Zhejiang Science and Technology Press.
21. Xu, G., and J. Zhang. 2014. *The third financial revolution of credit system: Financial evolution in view point of complexity sciences*. Beijing: China Finance Publishing House.
22. Yu, J. 2009. *Study on legal safeguard of circular economy*. Beijing: Law Press.
23. Zhang, Z. 2009. *Study on fiscal policy of circular economy development in hunan province*. Changsha: Hunan University Press.
24. State Environmental Protection Administration. Planning Guideline of Circular Economy Demonstration Area (Trial) [EB/OR]. http://www.mep.gov.cn/info/gw/huanfa/200312/t20031231_87885.htm.
25. Circular Economy Promotion Act of the People's Republic of China [EB/OR]. 2016. http://www.gov.cn/flfg/2008-08/29/content_1084355.htm.

26. China's circular economy network. From Concept to Action-main practices of building recycling-oriented society in Japan [EB/OR]. [2016–08–09]. http://www.chinace.org.cn/Content.asp?ID=10447&Class_ID=16.
27. Mao, J., J. Liang, and L. Xu. 2013. Service trend of mineral resources: Concept, connotation and topic. *Environmental Science & Technology* 36 (12): 195–201.

11 Sustainable Development Enterprise

11.1 Presentation of Core Issues

From the analysis in the previous chapters, it can be observed that a circular economy has a complex structure and its economic component has an enterprise as the minimum independent accounting and basic work unit. An enterprise is critical in achieving the ambitious goal of a circular economy. To clarify the basic requirements of a circular economy, an enterprise is very important to promote a circular economy; additionally, the enterprise must possess specific conditions and features to satisfy the overall objectives of the circular economy. For an enterprise to function in the social economy, the enterprise provides a product with a specific service function for the society or directly provides a specific service; thus, what is the performance of these products or services, and how is it evaluated? These are related to the survival of the enterprise; therefore, the key topics in this chapter are defining the requirements of a circular economy for an enterprise and determining how the product or service provided by an enterprise can be evaluated.

Beginning with the basic requirements of a circular economy for an enterprise, the concept of a sustainable development enterprise will be introduced and combined with the current environmental bottleneck factors confronted by enterprise development at this stage, the relationship between an environmentally friendly enterprise and a sustainable development enterprise will be explained. Then, focusing on the enterprise's core function of providing services to society combined with the product life cycle, an international standard is evaluated, and the basic method of enterprise-environment performance is discussed and evaluated.

J. Mao et al., *Circular Economy and Sustainable Development Enterprises*,
https://doi.org/10.1007/978-981-10-8524-6_11

11.2 What Is a Sustainable Development Enterprise?

11.2.1 What Are the Requirements of a Circular Economy for an Enterprise?

In the model framework of a circular economy shown in Fig. 6.1, the three activities of material production, product processing, and manufacturing, and waste recovery and regeneration consist of many related enterprises, and product use is not only the job objective of production and manufacturing but is also the premise of waste recovery and regeneration. There also exists an intrinsic connection to the enterprise's behavior; therefore, an enterprise is important in the operation of a circular economy.

For an enterprise's function in a material circular flow, as the necessary link in the material's artificial flow process, no matter if there is a "serial type," "parallel type," or other networking relationships among the enterprise and other enterprises, the material use efficiency in the entire artificial flow process will be influenced by the material utilization efficiency of that enterprise. The enterprise must provide its material utilization efficiency and embody the "3R" principle (i.e., reduction, reuse, and recycle) from different perspectives. This is primarily reflected in the highly efficient utilization of resources emphasized in and among each production department. The production department should establish substantial connections with the service and social consumption departments to maintain and extend product service life and to promote the waste recovery and recycling of the product after use. For the external enterprise, cost saving and emission reduction should be achieved to allow the entire industry to reduce its interference or influence on the external environment. To describe the environmental impact behavior of the enterprise more clearly, the enterprise causing less interference or influence on the human environment is called an environmentally friendly enterprise.

11.2.2 Which Essential Characteristics Should an Enterprise Possess?

The enterprise offers specific services for society and obtains certain economic benefits, which is the basic condition for the survival and development of an enterprise. During a resource environmental crisis, the enterprise possessing good environmental performance will become the enterprise that survives.

(I) Economic sustainability

Economic sustainability of an enterprise is embodied in their eco-environmental impact, which effects their economic development; furthermore, the sound development of an ecological environment is regarded as the basic condition of an enterprise's economic development. In the enterprise's economic development, the environmental friendliness of the product or service is regarded as the enterprise's competitiveness to promote resource use efficiency and service quality by means of technical innovation (including product variety, material selection, manufacturing technique, and service format) and the optimization of the inner link of material and energy among other enterprises, resulting in improved economic benefits. For example, an enterprise attracts customers by producing modular products convenient for the renewal of the product function or the replacement of parts damaged during use, or the enterprise gains customers via the close-to-life products made of environmentally friendly materials, gaining corresponding economic benefits.

To achieve economic sustainability, an enterprise should focus on the trend of national or regional economic development, understand their relevant industry, professional policy, and regulation, incorporate itself into a regional eco-industry system in a timely manner, and cooperate with other local enterprises to form an industrial cluster, thus gaining the corresponding regional advantages.

(II) Social sustainability

Satisfaction of social demands is one of the important objectives of an enterprise. Social sustainability of an enterprise means the enterprise should constantly meet the ever-growing and ever-changing social services requirements (e.g., the constant change in the performance requirements, appearance, carrying, and information storage of mobile communication devices) demanded by human beings. In this example, the mobile phone has been upgraded to incorporate a microcomputer device integrated with communication, video frequency, and other functions from the single "talking" device of its early stages. Second, by means of the produced product or provided service, the enterprise guides, meets, and constantly improves social consumption. For example, to ease traffic congestion and improve traffic conditions in Beijing, in 2013, the Beijing Traffic Committee installed 520 public bicycle rental locations to place 14,000 bicycles in seven districts and counties of Beijing, which provides bicycle access to four districts and counties, namely Dongcheng District, Chaoyang District, Fengtai District, and Shijingshan District and guides citizen to choose a green travel consumption pattern. In addition, the enterprise provides corresponding employment opportunities for society to promote the development of the labor force. Since the circular economy has developed, our country has developed a vast materials recovery and regeneration system and has provided a significant number of jobs for society. According to statistics, the number of people engaged in neighborhood services, repair, and other service industries had reached 754,000 in 2014.

(III) Environmental sustainability

Resource and environment are the basic conditions for enterprise development. The behavior of a sustainable development enterprise should be beneficial to the sustainable development of the environment. Environmental sustainability of an enterprise should be embodied in its operation to be beneficial for natural resource conservation and environmental waste emission reduction and should fully reflect a resource and environment protection strategy. For example, clean energy should be primarily selected to power equipment by an energy-using equipment manufacturing enterprise. There are very good cases, such as the solar water heater and solar car being developed in rapid succession in recent years. The service-type enterprise providing large-scale public service must greatly improve their service structure and operations pattern. For example, in recent years, Beijing has newly built subway routes to increase public transportation and to facilitate the substitution of a public transit system for private automobiles, concurrently, a number of bicycles are available to solve traffic congestion; thus, effectively improving environmental quality.

11.2.3 What Is a Sustainable Development Enterprise?

(I) Concept of a sustainable development enterprise

Due to the irreversibility of resources and environment conditions, reaching sustainable development is a lengthy process, and the enterprise will be requested to possess good environmental performance well into the future. Being "environment-friendly" is likely to become necessary for a successful enterprise. Therefore, the enterprise's development must not only gain social and economic benefits but also perform the mission of environmental construction. An enterprise that is beneficial for the sustainable development of the environment, economy, and society simultaneously is called a sustainable development enterprise. It is an enterprise that is guided by the "3R" principle of a circular economy to perform resource protection and environmental improvement-oriented process management of its product or service and to fundamentally achieve the "triple-win" objectives of economic, social, and environmental benefits.

It is not difficult to observe that an environmentally friendly enterprise is the initial and necessary stage of a sustainable development enterprise. It is the historic choice to break through the development bottleneck of the enterprise under the limitations of a resource and environmental crisis and is also an important measure to improve the resources and environmental conditions in the construction of a sustainable development enterprise. With the improvement of resources and the environment, an environmentally friendly enterprise at this stage will gradually transition to a substantive sustainable development enterprise.

(II) Connotation of a sustainable development enterprise

A sustainable development enterprise is embodied in the relationships within the enterprise itself and the relationship between the enterprise and the external environment. For the enterprise itself, the sustainable development enterprise is embodied in the circular flow relationship of the materials in each of its processes and among the processes; it is the basic representation of the circular economy at a microscopic level. It includes the following characteristics: The material lost in production is regarded as raw materials after being recycled for return to the original production process, the wastes produced in the production process are properly treated as raw materials or used to replace a portion of the raw materials to return to the material production flow, and the waste products recycled from society are reused. Enterprise behavior will be presented as providing an eco-friendly product or service with specific serviceability, including the overall improvement of the entire process from material selection, product manufacturing, and transportation of the product to the final user, delivery, product scrap recycling, and disposal. Within the relationship between an enterprise and the external environment, a sustainable development enterprise will focus on the integration of resources and environmental conditions within the region and actively form a communal relationship among resources (e.g., material, energy, and information) with other enterprises in the region that not only utilizes and saves other enterprises' waste resources but also provides its own unrecyclable wastes to other enterprises or departments for reuse or as garbage for disposal, to form a symbiotic eco-industry park or industrial cluster. Thus, by means of the eco-friendly construction of the enterprise itself, the coexistence of enterprise development and human sustainable development can be achieved.

(III) Relationship between an environmentally friendly enterprise and a sustainable development enterprise

An environmentally friendly enterprise grows specific to environmental improvement, and it is the enterprise whose environmental performance of its product or service is improved energetically on the basis of existing enterprises. Eco-friendly is regarded as the necessary attribute of enterprise development for its management and operation. It is not difficult to observe that an environmentally friendly enterprise is focused on environmental improvement, and it is the initial and necessary stage of a sustainable development enterprise. It is the historic choice to break through the development bottleneck of the enterprise under the limitations of the resource and environmental crisis and is also an important measure to improve the resources and environmental conditions in the construction of a sustainable development enterprise. With the improvement of resources and the environment, an environmentally friendly enterprise will gradually transition into a substantive sustainable development enterprise.

11.3 Method to Evaluate the Environmental Performance of an Enterprise

As the resource transformation entity, an enterprise is related to many raw materials, energy and power, manufacturing techniques, technologies, equipment, and infrastructure during the process of transforming raw materials into products. There is the exchange of material and energy between the enterprise and the external environment during enterprise operations, resulting in environmental impact. Effective and scientific methods are used to objectively understand these environmental impacts, so that measures can be taken to improve the enterprise's environmental performance.

Since 1997, evaluation methods for a product life cycle have been successively issued by the ISO to provide important scientific methods for the environmental performance of product assessment, specific services, and processes, and it becomes an important basis for determining the environmental performance of an enterprise as well.

According to the existing LCA standard, a product is used as an example in this section to specify the basic method for evaluating a product's environmental performance.

11.3.1 Overview

(I) Basic concept

A product is an important form for the enterprise to provide services for human beings. During the product life cycle, the necessary connection is established between the social economy and the external environment regarding material flow, energy flow, and other aspects. Therefore, we can determine the environmental impact level by analyzing the environmental impact caused during the product life cycle.

The life cycle described here refers to the entire process from obtaining raw material to product production, use, and waste disposal; it is the entire process of a product from "cradle-to-grave," and each stage is called a life cycle stage. For example, the life cycle of a car includes obtaining raw material (i.e., iron ore mining), forming steel products, glass, rubber, and other multiple industrial materials for preparing the various components of a car, processing and assembling components into the car, using the car after delivery, scrapping the car after several years of use, recovery, dismantling, and regeneration disposal.

Life cycle assessment (LCA) is a technique to evaluate environmental performance and potential environmental impact of products, and this technique is used to evaluate the potential environmental impact related to the input and output of a product system. The scope of evaluation involves the entire life cycle of a product. Through product LCA, the research result will be used to identify the improvement opportunities for product's environmental performance; provide the decision basis for the enterprise, government, and non-governmental organization for strategic planning and public policy making; choose the indicators to evaluate product environmental performance; prepare product environmental labeling; and promote marketing.

(II) Issued international standard

In 1993, the technical committee for environmental management was established by the International Organization for Standardization (ISO) to prepare the LCA international standard. In 1997, the principle and framework of the LCA standard were issued, and then ISO 14040-14049 (nearly ten international standards) were successively released, as follows:

ISO 14040: 1997 Environmental management—Life cycle assessment—Principles and framework are replaced by ISO 14040: 2006 Environmental management—Life cycle assessment—Principles and Framework.

ISO 14041: 1998 Environmental management—Life cycle assessment—Goal and Scope Definition and Inventory Analysis.

ISO 14042: 2000 Environmental management—Life cycle assessment—Life Cycle Impact Assessment.

ISO 14043: 2000 Environmental management—Life cycle assessment—Life Cycle Impact Interpretation.

ISO 14044: 2006 Environmental management—Life cycle assessment—Requirements and guidelines.

ISO/TR 14047: 2012 Environmental management—Life cycle assessment—Illustrative examples on how to apply ISO 14044 to impact assessment situations. The original ISO/TR 14047: 2003 Environmental management—Life cycle impact assessment—Examples of application of ISO 14042 is replaced.

ISO 14048: 2002 Environmental management—Life cycle impact assessment—Data documentation format.

ISO/TR 14049: 2012 Environmental management—Life cycle assessment—Illustrative examples on how to apply ISO 14044 to goal and scope definition and inventory analysis. Replaces the original standard ISO/TR 14049: 2000 Examples of application of ISO 14041 for goal and scope definition and inventory analysis.

(III) LCA framework

Product LCA includes the following four basic elements: the definition of research purpose and scope, inventory analysis, environmental impact assessment, and explanation of research results, as shown in Fig. 11.1.

11.3.2 Definition of Purpose and Scope

(I) Definition of research purpose

To implement product LCA, first, clarify the purpose of the research. Clearly specify the application intention of the LCA study, the cause of the study and its application object. For example, inform consumers of the environmental performance of a product or compare two products with same service function to make the choice beneficial for environmental protection. The service object of the

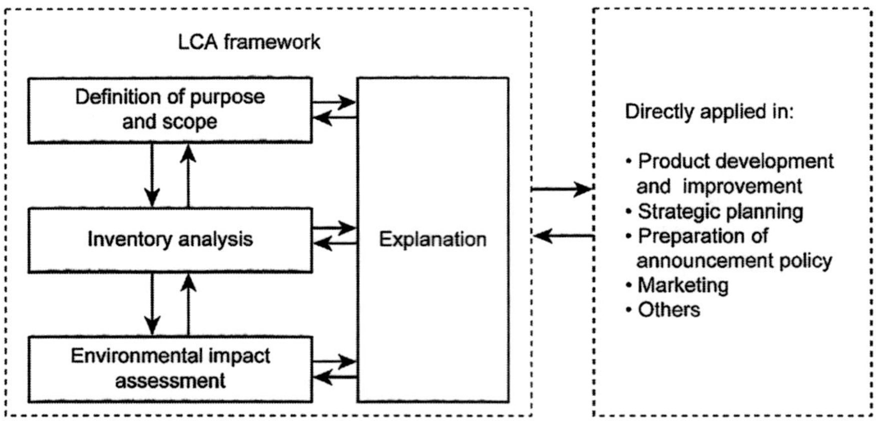

Fig. 11.1 Composition of product LCA. *Source* ISO 14040

research result can be an enterprise manager, functional government department, product consumer, or scientific research institution.

(II) Definition of research scope

To ensure consistency among the study range, depth, level of detail, and research purpose, the scope of the LCA study must be defined, including the following:

1. Product function, functional unit, and reference flow

The function of the product system refers to the service performance of product, e.g., tissue is used to dry an arm and a light bulb is used for lighting. When a product has multiple service functions or can provide multiple services, a specific function for specific LCA research must be chosen based on the purpose of the LCA study.

A function unit (FU) refers to the unit selected to quantitatively describe the service performance provided by the product system. For example, a 10,000 m^2 wall is painted and maintained for five years, which can be used as the function unit provided by the painting product. For another example, in a transportation system, "10,000 people are transported from A to B every day for ten years of continuous use," which can be used as the function unit of the service system. Generally, the function unit is the composite unit composed of many physical units; this is entirely different from the traditional physical unit.

Reference flow is the parameter to measure the output quantity of each process or production phase in the product system to achieve the specific amount of service function represented by the function unit. For example, when a "20 m^2 wall is painted and maintained for ten years" is defined as the service function unit of a specific painting product, if the painting performance is that 1 L of paint is sufficient to cover 8.7 m^2 of wall for five years, the "20 m^2 wall is painted and maintained for ten years" can be achieved, and the quantity of paint required can be calculated as follows:

$$\frac{20 \times 10}{8.7 \times 5} = 4.6\ \text{L}$$

Put differently, to complete the specific quantity of service (i.e., "20 m^2 wall is painted and maintained for ten years"), 4.6 L of paint is required, which is the reference flow corresponding to the paint product system.

2. Product system and system boundary

The product system is the set of processes for the entire product life cycle. The system boundary is the interface between the product system and its external environment. Generally, the composition of a product system depends on the research purpose of the LCA. When the same product has different research purposes, its defined product system is also different. For example, when a specific structural steel product is used to compare the energy conservation level between iron ore steelmaking and steel scrap resource steelmaking, LCA research would be related to two types of product systems: The semi-life cycle primary steel product system composed of iron ore mining, blast furnace ironmaking, converter steelmaking, steel rolling, and other major process, and the semi-life cycle steel scrap regeneration product system composed of scrap steel recycling, electric arc furnace steelmaking, steel rolling, and other processes.

3. Categories of environmental impact

When LCA research is implemented, the environmental impact categories generated in the product must be chosen. In ISO 14040 and ISO 14043, the environmental impact is divided into three impact types: ecosystem, human health, and natural resources. In application, the environmental impact is subdivided into global climate change, ozone depletion, acid deposition, biodiversity reduction, and other subclasses. When a specific LCA is implemented, generally, one or more environmental impact categories are chosen for study. For example, to determine which of the original refrigerants [i.e., HFC-134a and R12 (CFC-12)] is more environmentally friendly, consider that the refrigerant primarily influences the composition of the atmospheric material in the ozone sphere; therefore, ozone depletion is selected as the category of environmental impact for the refrigerant LCA study.

4. Data quality requirement

In the LCA research scope, the data quality used in the LCA must be defined according to the purpose of the LCA study. This includes data-related time spans, regional breadth, technical coverage, data accuracy, integrity and representativeness, data source, and typicality. For example, for a specific LCA research of an upgraded manufacturing technique, the data used in the study are required to be from the actual production of the enterprise.

(III) Several examples
1. Example 1: Identification of product function and the function unit

To understand the special concept of the function unit more clearly, Table 11.1 shows examples of several products and compares the differences in the product performance and the function unit.

2. Example 2: Example of a product system

As previously mentioned, a product system is the set of processes for the entire product life cycle. It includes all activities from raw material mining to final waste disposal, as shown in Fig. 11.2.

11.3.3 Inventory Analysis

(I) Inventory analysis steps

Inventory analysis is the process of preparing the input and output lists specific to each process in the product system and calculating the data corresponding to its functional unit. The basic steps are shown in Fig. 11.3.

It can be observed from Fig. 11.3 that the inventory analysis in life cycle assessment is primarily related to data collection and calculation.

Table 11.1 Example for identifying product function and function unit

Product	Bulb	Bottle	Tissue
Function	Lighting Heating Others	Containing drinks Easy to carry Product image design Others	Wiping arms and hands Cleaning dirt Others
Product functions for specific LCA	Providing indoor illumination	Containing drinks	Wiping arms and hands
Product performance	Illuminance 100 lx, service life 1000 h	Capacity/pcs. 0.5 L	1 pcs. tissue for wiping 1 hand
Function unit	Achieving illuminance 300 lx, service life 50,000 h	Containing 50,000 L of drinks	Wiping 1000 pairs of hands
Reference flow	150 bulbs with 100 lx of illuminance and 1000 h of service life	100,000 beverage bottles with the capacity of 0.5 L	2000 pcs. of tissue

Source ISO 14049

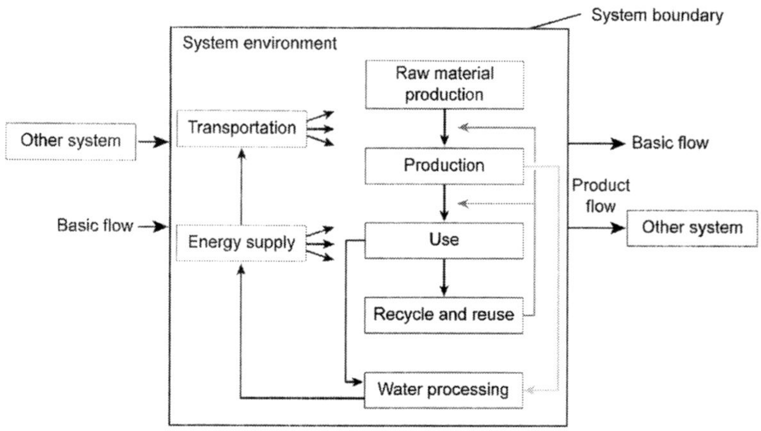

Fig. 11.2 Sketch map of product system. *Source* ISO 14040

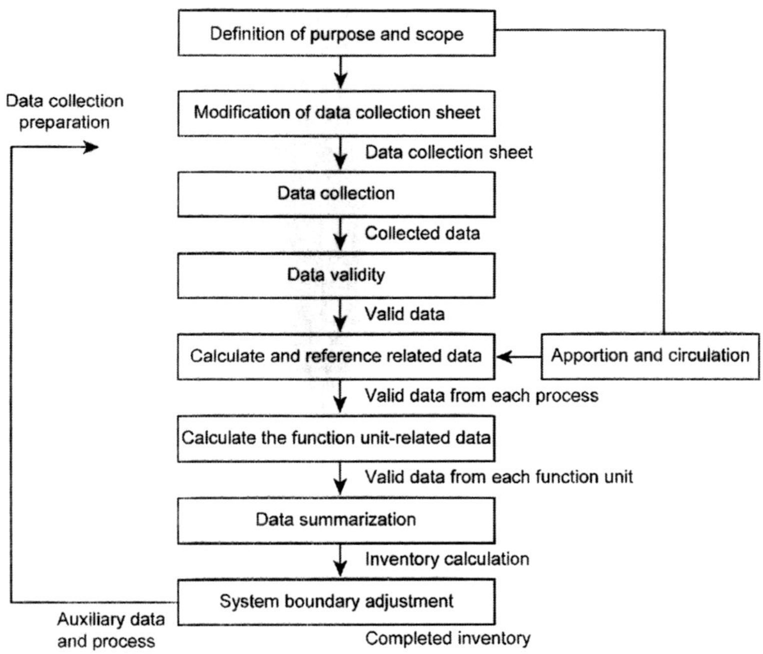

Fig. 11.3 Basic steps of inventory analysis. *Source* ISO 14041

(II) Data collection

To obtain all the data demanded for a product LCA, generally, perform the following steps.

First, decompose each step in each life cycle phase in the product system listed in the research scope to form a technique flowchart at each relevant enterprise level. Specifically, draw the process flow diagram for the different manufacturing shops and production processes in each enterprise with sufficient detail to independently collect actual operational data, and then, draw the function of each process in the product life cycle and the relations among them. Figure 11.4 is the industrial flow diagram of a specific steel plant and the process flow diagram of the rolling production line.

Second, describe the function of each process (including the process name) in the product life cycle and list the type of data collected and their units of measurement, as shown in Table 11.2, to perform data collection specific to every process.

The process is the basic unit to collect data. When data are collected specific to each process, we can list the types of required data, as shown in Fig. 11.5. Then, collect each piece of related data. If data are not collected in the production field, a data collection guideline should be prepared to describe the data collection and counting method, and the handling method of the possible problem on site (e.g., data shortage or significant deviation), to guide field staff to reference and collect the required data.

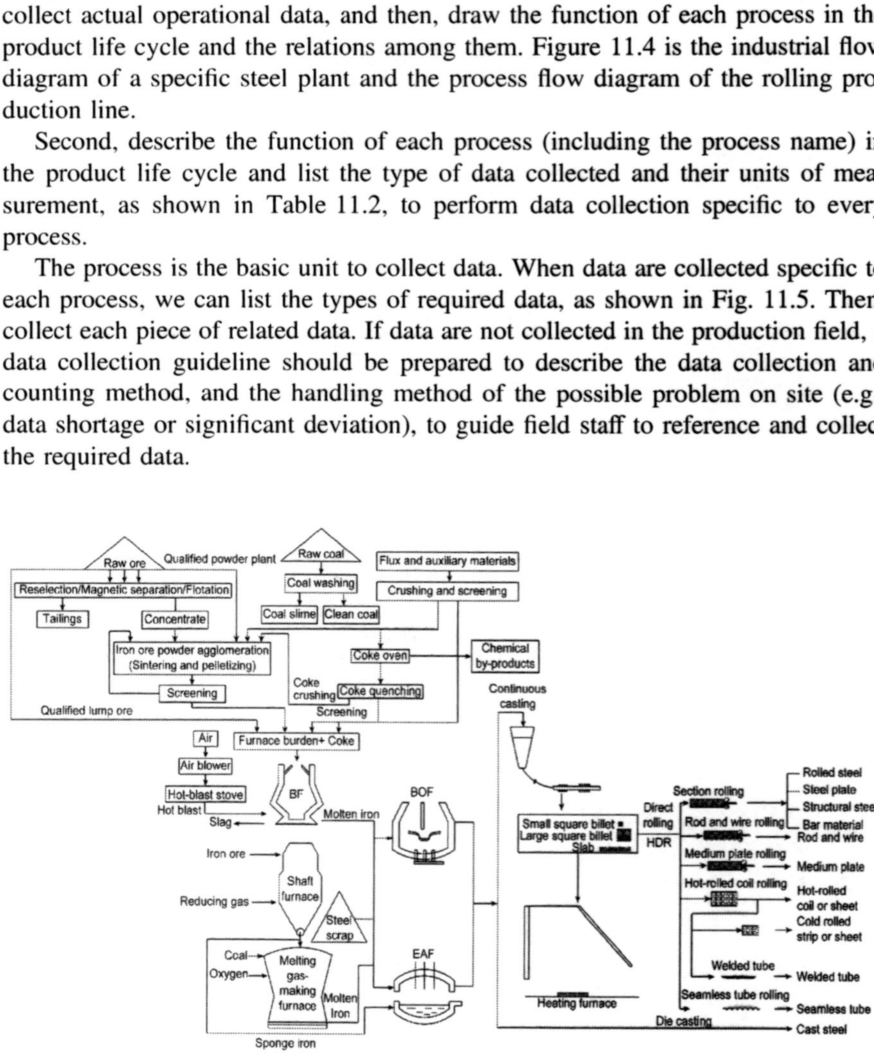

Fig. 11.4 Technological process of a certain steel manufacture process. *Source* Wang [1]

Table 11.2 Example of a process data collection table

Preparer		Process name		Process number	
Collection time		Starting and ending time		Collection location	
Process description					
	Unit	Quantity	Description of sampling method	Data source	Remarks
Material input					
Energy input					Gasoline, kerosene
Water consumption					Surface water, underground water
Product					
Exhaust gas					CO_x, NO_x, SO_x
Wastewater					BOD, COD
Solid waste					Hazardous waste, slag

Source ISO 14049

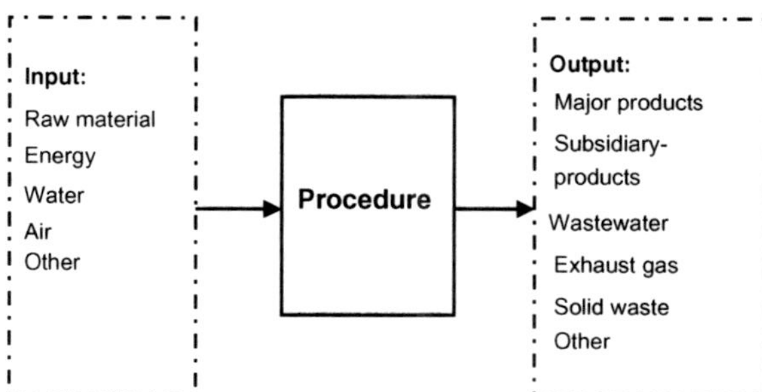

Fig. 11.5 Basic unit of inventory analysis

(III) Data calculation

The purpose of data calculation is to obtain the data corresponding to the function unit defined within the range of the study. The collected data are used to calculate the data corresponding to the reference flow in each process, and then, the data are converted to correspond to the function unit of the product system.

Generally, data calculation should consider the physical relation among data (e.g., material transformation relationship, transformation, and distribution relationship). For similar material or sources of energy, use the same unit of measurement. The source of energy is calculated based on standard coal. For unavailable data, provide the values via counting, estimation or another method, and indicate the method used.

Table 11.3 Example of an LCA data settlement result for the energy consumption of a glass bottle life cycle

Energy consumption process	Electric power (%)	Heating power (%)
Raw material mining and smelting	0.1	2.6
Glass production	4.5	14.2
Cleaning and filling	64.4	61.4
Using (refrigeration)	15.9	–
Recovery (cleaning)	0.1	–
Waste treatment	–	–
Label making	4.4	8.8
Bottle cap (the entire life cycle)	10.2	10.5
Casing	0.5	2.5
Distribution	–	–
Transportation	–	–
Total quantity	100	100
Total quantity (kW h or MJ)	78 kW h	750 MJ

Source ISO 14049

In the example in Table 11.1, when a beverage bottle is in volume production, according to the field data from its forming process, when 10,000 0.5 L beverage bottles are formed, 5 MJ of heat would be consumed; therefore, when 100,000 pcs. (relative to the reference flow) of beverage bottles are formed, 50 MJ of heat would be consumed. The input/output data of each process corresponding to 100,000 beverage bottles is calculated, and finally, the data of the entire life cycle corresponding to 100,000 beverage bottles are summed. As the 100,000 of 0.5 L beverage bottles correspond to the function unit of the LCA, the data are the data of the beverage bottle life cycle corresponding to the function unit, namely the quantity of heat consumed to contain 50,000 L of beverage. The result of the data summarization can be converted into an absolute value or a relative value; however, within the latter it is easier to observe the distribution difference of the environmental load in each element, as shown in Table 11.3.

An example of the energy consumption of a glass bottle life cycle is provided here, while the actual inventory results for an LCA will be related to different types of data (i.e., material, energy, and noise) and should be collected into the table as the result of an inventory analysis.

11.3.4 Environmental Impact Assessment

A product life cycle impact assessment (LCIA) is applied to evaluate the environmental impact level of the product (or service) system based on the result of the inventory analysis of the product life cycle. The environmental impact assessment is related to the type and indicator selection of environmental impact, classification of inventory analysis results and characterization.

(I) Selection of environmental impact indicator and characteristic model

Based on the types of environmental impact selected in the definition process of the previous study range, and based on the two relation models (also called the characteristic model) between each material causing environmental impact and the formed environmental impact effect established by the international research institution, determine the indicator representing the level of environmental impact, namely the environmental impact category indicator. For example, due to climatic variation, the characterization model is selected from the relation models between greenhouse gas emission and climatic change established by the Intergovernmental Panel on Climate Change (IPCC). For such environmental impact of climatic change, infrared intensity is used as the indicator to represent the climatic change by the IPCC, and global warming potential (GWP) is used to reflect the climatic changeability of the different greenhouse gases. Table 11.4 shows the relationship among inventory analysis results, environmental impact indicators, and characteristic models. Other types of environmental impacts are also researched by the special international organization to build the corresponding models and characteristic indexes.

(II) Classification of inventory analysis results

Based on the environmental impact categories selected in the definition of the range of the study, the results of the life cycle inventory analysis are classified into environmental impact categories. For example, in a specific LCA, climatic change, and acid rain are selected as the environmental impact categories, and the amounts of CO_2, CH_4, SO_2, NO_2, and other pollutants are obtained in the analysis. In this case, the discharge of CO_2 and CH_4 should be classified as climatic change, while the discharge of SO_2 and NO_2 should be classified as acid rain.

Table 11.4 Example of terms in the determination process of environmental impact categories and characteristic index

Term	Example
Inventory analysis LCI result	Greenhouse gases
Environmental impact categories	Climatic change
Characterization model	IPCC model
Index of environmental impact categories	Infrared intensity (W/m^2)
Characterization coefficient	GWP of various greenhouse gases ($kg\ CO_2$ equivalent/kg gas)
Index calculation result	Carbon dioxide equivalent amount (kg)

Source ISO/TR 14047

(III) Inventory result characterization

One type of environmental impact category may be caused by many environmental pollutants, e.g., global warming might be caused by CO_2, CH_4, N_2O, and other greenhouse gases. To uniformly represent the environmental impact ability of different pollutants more effectively, a typical substance is selected to convert the environmental impact energy of other pollutants by building the proportional relation of the quantity between other pollutants and the material causing the same environmental impact as the typical substance. The proportional relation of the quantity between a specific material and a typical substance under same environmental impact is called the characterization coefficient of the material. The characterization system is usually determined by a special research institution during the building of a characteristic model, as described in Sect. 4 of Chap. 4. For example, in the climatic change problem, the IPCC defines the climatic changeability of CO_2 within 100 years to be 1, and other substances are converted into the equivalent amount of CO_2 based on their climatic change abilities. This converted coefficient is called a characterization coefficient. For example, the CH_4 GWP is 25 and the CO GWP is 298. The different inventory analysis results are converted into the corresponding values of environmental impact via various environmental impact characteristic coefficients, which is called environmental impact characterization. The process can be divided into two steps. First, the characterization coefficient is selected and used to convert the assigned LCI results into the unified unit (i.e., the amount of all greenhouse gas emissions is expressed as a "CO_2 equivalent"), and the calculation results with the unified unit are collected. Second, the calculation results of various environmental impacts are summed to obtain the quantitative result of the life cycle environmental impact.

(IV) Example

The light bulb in Table 11.1 continues to be used as an example.

If the discharges of greenhouse gases CO_2, CH_4, and N_2O are 50, 30, and 10 kg, respectively, in the life cycle of 10,000 bulbs with 100 lx illuminance and 1000 h of service life, what is the impact level of climatic change demanded to achieve 300 lx 50,000 h of lighting?

Solution:

Different greenhouse gases have different climatic change environmental impact characterization coefficients, namely, their GWP. Considering that the GWP values of CO_2, CH_4, and N_2O are 1, 25, and 298 kg CO_2 equivalent/kg, respectively, the GWP per 10,000 pcs. of 100 lx, 1000 h bulbs is as follows:

$$50 \times 1 + 30 \times 25 + 10 \times 298 = 3780\,(\mathrm{kg\,CO_2\,equivalent}/10{,}000\,\mathrm{bulbs})$$

Based on the estimated results in Table 11.1, the reference flow corresponding to the process of achieving 300 lx, 50,000 h lighting is 150 bulbs, and the environmental impact value corresponding to the reference flow is as follows:

$$3780/10,000 \times 150 = 56.7\,\mathrm{kg\,CO_2\,equivalent}$$

The environmental impact value corresponding to the reference flow is the environmental impact generated by achieving the designated function unit, and it can be expressed as the quantity of the function unit (FU), as follows:

$$56.7\,\mathrm{kg\,CO_2\,equivalent/FU}$$

In addition, there are some optional elements, as specified in the following specialized courses or referenced professional books.

11.3.5 Discussion and Explanation

The explanation and discussion stage of the product life cycle is phase IV of the LCA, and it is used to explain and discuss the results of the LCA. In this phase, the researcher should identify the various influential factors that might affect the LCA research results and analyze the possible influence of these factors on the LCA research results to form conclusions and the report of the LCA research.

(I) Example of identifying important matters

During the process of identifying LCA influential factors, focus on whether the research methods satisfy the requirement of the research purpose. These research methods include the selected product system (e.g., whether the established product system provides the expected service performance appropriately), the composition of each life cycle stage (e.g., the composition of each process), the type of inventory data (e.g., energy, environmental emission, if a specific type of data is lost, e.g., noise, is it possible to obtain different LCA research conclusions) and the environmental impact category (e.g., utilization of mineral resources, climatic change). These factors are called the important matters of environmental impact.

(II) Data evaluation and inspection

LCA research results and conclusions are directly influenced by the data quality used. To ensure objective and valid research results, the data used in the study must be inspected and evaluated, including the following aspects:

1. Completeness check

A completeness check is used to confirm the availability and completeness of the information or data required in the LCA study, as shown in Table 11.5. Data found to be missing during the inspection can be remedied via actual measurement or estimation, and unnecessary data can be ignored, but its reasons for omission should be recorded.

2. Sensitivity check

The sensitivity check is used to evaluate the reliability of the LCA research results and analyzes whether the LCA research results are influenced by data quality, calculation methods, a characteristic model of environmental impact or other various reasons, and the impact of their influence. Table 11.6 shows an example of the sensitivity check of the characterization coefficient.

Table 11.6 shows that the value of the characterization coefficient has 28.6% sensitivity to the research results; thus, it is the LCA sensitive parameter.

3. Consistency check

Verify whether the assumptions, method and data used in the LCA research are consistent with the content defined by the research purpose and scope, as shown in Table 11.7.

Table 11.5 Example of a data completeness check

Process	Optional item	Is the data complete?	Required improvement action
Materials production	×	Yes	–
Energy supply	×	Yes	–
Transportation	×	?	Checklist
Processing	×	No	Checklist
Packaging	×	Yes	–
Use	×	?	Comparison with others and data supplement

Source ISO 14047

Table 11.6 Example of the sensitivity check of the characterization coefficient

GWP data input	Method A	Method B	Difference
GWP score = 100 CO_2 equivalent	2800	3200	400
GWP score = 500 CO_2 equivalent	3600	3400	−200
Absolute deviation	+800	+200	600
Relative deviation (%)	+28.6	+6.25	Significant deviation
Sensibility (%)	28.6	6.25	–

Source ISO 14047

Table 11.7 Example of a consistency check of the data

Check	Method A		Method B		Comparison	Action
Data source	Literature	Available	Original data	Available	Consistent	No
Data accuracy	Good	Available	Poor	Inconsistent with the range of study	Inconsistent	B is implemented again
Time limit of data	2 years	Available	3 years	Available	Consistent	No
Technical coverage	Advanced technology	Available	Present technology	Available	Inconsistent	Conform to the requirement of research purpose, no action
Time scale	Recently	Available	Actual value	Available	Consistent	No
Region scale	Europe	Available	USA	Available	Consistent	No

Source ISO 14047

11.3.6 Writing the Research Report

Writing an LCA research report produces the overall summary of the LCA research process and results. An LCA research report should show just, complete and accurate research results to users, including the following contents:

(1) Research purpose: cause, application purpose, and service object;
(2) Research scope: range of product system, product function, function unit, system boundary, data category, data quality requirement,, and environmental impact category;
(3) Inventory analysis: data collection method, process description, literature resources, data calculation method, and data efficiency analysis;
(4) Impact assessment: environmental impact category, category indicator, characteristic model, and LCI results in classification and calculation;
(5) Explanation and discussion: important matter analysis of results, research quality assessment, conclusions, and suggestions.

Each element influencing the research results should be fairly, completely, and accurately reported to the users, so that readers can understand the complexity of the research and the possible problems.

Evaluating the environmental impact level of an enterprise's products is an important method to understand the environmental performance of their behavior. The implementation of an eco-friendly enterprise is necessary for the enterprise to move toward becoming a sustainable development enterprise and is also a necessary stage to realizing of global sustainable development, therefore the enterprises should been encouraged to perform environmental evaluations.

Classroom Discussions and Assignments

Topic Discussion

The product LCA method cannot only evaluate the environmental performance specific to a certain product or service but can also evaluate a specific production process or manufacturing technique. Please choose a familiar product or service and provide an outline for an environmental impact assessment.

Homework and Presentation in Class

By means of previous class discussions, homework, and presentations in class, students have learned the skills of topic selection, the basic thinking and methods of research of a specific topic, and can summarize it into a PowerPoint team report on the subject. However, in many cases, it is difficult for interested readers to personally attend the PowerPoint presentation; thus, the research must be presented as published papers. Paper writing is another important method to present the research engineering and results to readers. For this class, students are requested to present their scientific research in the form of a paper. The task of this lesson is to read an LCA research paper on the product or service selected in the class discussion and presented to the class as a PowerPoint presentation by the group, and summarizes it into a 1500- to 2000-word essay. To present the research status of a paper, include the topic selection process and the understanding and viewpoints of the paper. The grading scale is shown in Table 11.8.

Table 11.8 Grading of the group report in class

PowerPoint report quality in the class (weight 0.5)			Essay writing (weight 0.5)			Total points
Process of topic selection (Maximum 10 points)	Research basis (Maximum 10 points)	Result and conclusion (Maximum 10 points)	Topic selection and viewpoints (Maximum 10 points)	Paper summary (Maximum 10 points)	Expression and specification (Maximum 10 points)	

Recommended and Referenced Literatures

1. Wang Xiaoliu. 2013. *Metallurgy of iron and steel* (Ironmaking Part). Beijing: Metallurgical Industry Press.
2. Luo, Bangsheng, and Anzhou Li. 2005. *Corporate environmental management*. Chongqing: Chongqing University Press.
3. Ge, Chazhong, and Hongqiang Jiang. 2010. *Industrial pollution control and environment friendly enterprise*. Beijing: China Environmental Science Press.
4. Pollution Control Division of Environmental Protection Department, and Environmental Development Centre of Environmental Protection Department. 2009. *Work manual of the establishment of national environment friendly enterprise*. Beijing: China Environmental Science Press.
5. Li, Jingjiang. 2006. *Enterprise' green operation-the only way of sustainable development*. Beijing: Tsinghua University Press.
6. Chu, Aili. 2010. *Circular economy and enterprise sustainable development*. Beijing: Beijing Normal University Press.
7. Lu, Zhong-wu. 2009. *Foundation of industrial ecology*. Beijing: Science Press.
8. Kannan, R., K.C. Leong, and R. Osman et al. 2007. Life cycle energy, emissions and coast inventory of power generation technologies in Singapore. *Renewable & Sustainable Energy Reviews* 11:702–715.
9. Yang, Jianxin, et al. 2002. *Product life cycle assessment method and application*. Beijing: China Meteorological Press.
10. Deng, Nansheng, and Xiaobing Wang. 2003. *Life cycle assessment*. Beijing: Chemical Industry Press.
11. Chen, Sha, and Zunwen Liu. 2014. *Life cycle assessment and Type-III environment label certification*. Beijing: Chinese Quality Inspection Publishing House.
12. Yu, Suiran, and Jing Tao. 2012. *Design and assessment of product full life circle*. Beijing: Science Press.

12 Eco-Design

12.1 Presentation of Core Issues

Previously, the type of property an enterprise should possess was specified from two aspects, namely the requirement of a circular economy for an enterprise and the role of an enterprise to achieve a circular economy. The enterprise should meet the specific demands of society, belonging to social performance. To achieve this service target, the enterprise transforms natural resources into a product with a specific service function, and in this type of production, it not only must input a significant amount of material resources, but the energy driving the material transformation, various production equipment and facilities, manpower and production space, which means that the value of the material is increased during the production process, reflecting the economic performance of the enterprise. The material placed into production is ultimately from the natural environment, and furthermore, all materials incapable of being put into the production process can be transformed into final products or can be reused by other departments. The material not used in the human socio-economic system will be discharged into the environmental system to be a pollutant in the environment, resulting in an impact on natural resources and the environmental system, so that the enterprise has environmental performance. On the basis of existing enterprise performance combined with the current resource and environmental problems, and specific to the stage of this type of problem, Chap. 11 emphatically describes the environmental performance of improvement-oriented environmentally friendly enterprises and analyzes the basic method for evaluating the environmental performance of the products manufactured by the enterprise from the perspective of providing services to society via the products, namely life cycle assessment (LCA).

There is no doubt that once a product is placed into production, it will have a specific quantity of environmental impact. Put differently, a specific quantity of environmental impact will be inevitable corresponding to the product's (or the service provided by such a product) life cycle process. Therefore, does a method

J. Mao et al., *Circular Economy and Sustainable Development Enterprises*,
https://doi.org/10.1007/978-981-10-8524-6_12

exist to prevent, avoid, or reduce the possible significant environmental impact of a product after it is placed into production before the product (or service) is formed? If so, what is the method?

All industrial products begin at their design. Put differently, design is the precondition to placing a product into production. If the possible environmental impact of a future product can be predicted effectively and prevented actively in the design stage, it may be possible to avoid the environmental impact from the product's origin. The design of the product and its technique are the preconditions of the product production; if there is a poor design, it is difficult to achieve the eco-friendly objective of the enterprise. From the perspective of enterprise production, eco-design is also indispensable to achieve the enterprise's sustainable development. Therefore, the methods to avoid significant environmental impact during design will be discussed in this chapter. It includes the concept and the major technical measures of eco-design.

12.2 What Is an Eco-Design?

12.2.1 What Activities and Influential Factors Should Be Considered in an Eco-Design?

Design is the preparatory process of a product. During a product's design stage, the traditional designer might consider the following three basic factors:

(1) Facing customers—The designer must consider how to achieve the demanded service function to meet the users' needs.
(2) Facing the manufacturing enterprise—The designer must determine the method to present the demanded services by means of production equipment, processes, and other basic conditions (i.e., the method to produce the product with a specific function).
(3) Facing other economic and industry departments—The designer must determine the method to obtain various materials and sources of energy demanded by production and the method to send the transformed product or service to users.

Each above consideration is further influenced by many social, human and productive factors, including the following:

(1) To meet consumer demand—This includes product performance, feeling, price, consumer preference, convenient maintenance/repair, and other factors. Different consumer demands require the design of different products. For example, in the selection of a vehicle, some might prefer a colorful and luxurious off-road vehicle with greater power, while some might prefer an economical and practical, portable and simple scooter.

(2) In enterprise production—It must be evaluated whether the enterprise's existing production equipment, manufacturing technique, and infrastructure possess the basic conditions to complete the expected product manufacturing process. Are there sufficient manpower and financial resources? Which new facility or process must be supplemented? Do the manufactured products conform to the existing laws, regulations, and design standards? For example, a fuel car is only produced in a specific automobile factory traditionally. Because the current market is increasingly demanding electric cars, the enterprise can consider the expansion of its business to produce electric cars. In this case, it must improve or increase its manufacturing shops for electric cars, especially to change the infrastructure of the automobile engine.

(3) Regarding the relationship between other economic and industry sectors—It must be determined whether the raw materials, auxiliary materials, and power facilities of the enterprise are convenient for purchase. Are they convenient to store? Where can they be purchased? What is the price? What is the status of the products' consumer group? What is the space-time range of marketing? Is the product easy to transport? What type of packaging and means of transportation should be used to send it to users? Compared with other similar products, what competitive advantage does it possess? How can adverse factors be avoided and improved?

12.2.2 What Type of Design Can Be Called an Eco-Design?

For the majority of products, their design is obtained through improvement of an original product combined with the specific needs of users. In the improvement process, the newly designed product must always possess some new features that are unavailable in the original product. In application, the design of a product possessing new features is called the design for X (DfX). The new features to be added are represented by X, where X might be the expansion of the product's usable range, easier to carry, more convenient for parts replacement, or more beautiful. For example, the television set evolved into an LCD ultra-thin model from the "slicked-back" that was popular in the 1980s, bringing clearer visual effect and greatly reducing floor space.

When the new features of the products are to improve the products' environmental performance, the design is called design for environment (DfE). This is product design for environmental improvement, which is also called eco-design or green design.

It can be seen that eco-design has an ecological priority. The environmental factor is incorporated into the design, and in the product design and its entire life cycle, the industry-related eco-environmental problems are comprehensively considered to design a new product that is environmentally friendly and can meet people's demands. The environmental factor must be considered in all stages of the product development by the eco-design to reduce the environmental impact from the entire life cycle of the product, to develop a more sustainable production and consumption system. It not only embodies the social value but also protects natural value, promoting harmonious development between human beings and nature.

12.2.3 What Are Its Differences from the Traditional Design?

Compared with traditional design, there are many differences in eco-design's service target, scope, and team, including the following:

(1) Different service target—The traditional product design serves the product manufacturing enterprise, focusing on the design of the product service performance and its production process. Eco-design emphasizes that product environmental performance should be integrated into product design, which not only serves the product manufacturing enterprise but also assumes the duty and responsibility of protecting the environment.
(2) Different scopes of design to be considered—This primarily involves the activities and influential factors to be considered. The life-cycle stage of product manufacturing is primarily involved in the traditional product design. While the scope of eco-design is expanded to the entire life cycle of the product, it not only must consider the product processing and manufacturing department but must also further consider material mining, product use, waste disposal, and other life cycle activities as shown in Fig. 12.1. The type of impact the entire life cycle of the product causes must be fully considered. How can these impacts be reduced? For this purpose, in product design, whether the materials of the product are full of resources, non-toxic, and harmless must be considered. Are there toxic and hazardous gases to pollute the environment during the technology and production process of the transformation of a substance from resources to products? What about the energy consumption in the product's use? Is the product's operation/maintenance more convenient? Is it

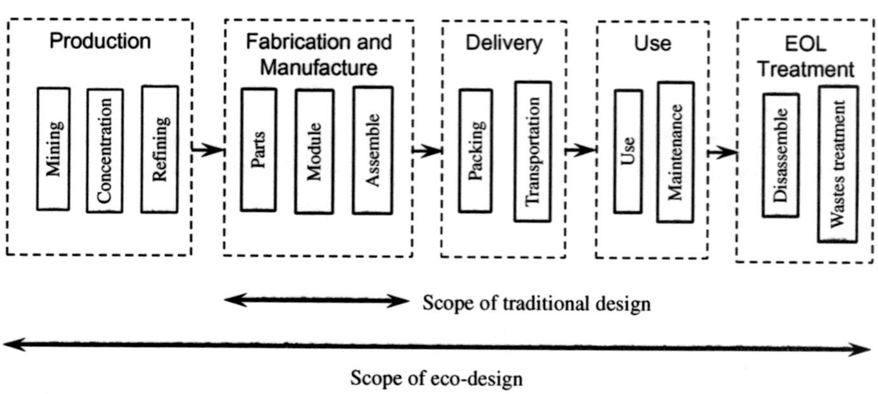

Fig. 12.1 Difference of the consideration scope between eco-design and traditional design

suitable for component renewal? Is it easy to dismantle after product scrapping? Are its wastes recycled? What about the harmless treatment and environmental emission of the waste? It can be seen that compared with traditional design, many factors that consider exterior ecological resources and the environment are added and stressed in an eco-design.

(3) The factors of resources and environment are integrated into the factors to be considered; eco-design should be consistent with environmental improvement, and it not only must consider global and national environmental management matters, e.g., global warming and destruction of the ecological environment, but must also consider regional and local small targets, e.g., eutrophication, toxic, and harmful substances exceeding standards in specific areas. As there is a significant difference in the resources and environmental status of different countries and regions, the environmental improvement goal in different areas is also different. The goal of local environmental improvement must be integrated into the improvement goals of the product environmental performance based on the local conditions. For example, in areas with sufficient oil resources, oil products should be selected as the power source for production or product use in product design, while in an area where a specific pollutant significantly exceeds standards, in the design, another material or service form should be used as a substitute for the services provided by the material.

(4) Different professional structure of the design team—The traditional product design team primarily consists of technical designers. In eco-design, as the product environmental performance must be considered and the design scope is more extensive, the design team should include experts in the environment, raw material purchase, packaging, marketing, product life cycle planning, and finance and should include the supplier of raw materials, product consumers, and other representative staff.

12.3 What Technical Measures Can Be Taken to Reduce Environmental Impact?

Based on the previous analysis, in product eco-design, the focus is on the raw materials of the product, the manufacturing technique demanded by the material transformation, the method to deliver products to final users, the method to maintain the product, disposal after use, and other basic elements. Several typical elements are selected, and the possible technical measures to improve their environmental performance are discussed below.

12.3.1 Material Selection

Product design must consider whether the materials in the product are full of resources, non-toxic, and harmless. Will toxic and hazardous gases pollute the environment during the technology and production process for the transformation of substances from resources to products?

(I) Factors to be considered in material selection

Industrial material is the material basis of a product; therefore, material selection is one of the important factors to be considered in the product design stage. When industrial materials are selected, the basic requirement of the product service performance for materials in conventional design must be satisfied, meaning that the selected material should possess the physical and chemical properties demanded for the proposed production of the product (e.g., intensity, electric conductivity, refractive index, solubility, light sensitivity, and reactivity) and be simple to obtain. What about its economic cost? Conversely, the requirement of material's ecological environment performance under the concept of eco-design includes whether the material is a scarce material. Is its supply limited by regional or monopolistic property? Are material processing and production safe? Does the required processing technology involve a hazardous substance? Is the material performance safe and stable? What is its cost?

(II) Eco-material

To represent the eco-environmental performance of industrial materials more effectively, the concept of eco-material is proposed. Eco-material is a type of material that consumes fewer natural resources, causes less environmental impact, and is slightly restricted by environmental regulations for mining and use procedures. The assessment of a material's ecological performance can be conducted from seven aspects of the supply of natural resource, material regenerability, and environmental impact. For each aspect, some specific scores are defined to quantitatively rank the related property, for instance, to estimate the richness of resource supply through resource depletion time, and define 1 as less than 10 years, while 2 as 10–20 years, 3 as 20–30, and 10 more than 100 years.

For a specific material, different properties may have different levels, and different properties are not on the same dimension. To see the eco-environmental performance of the materials more intuitively, based on the above seven aspects of evaluation indexes, its performance can be evaluated through a diagram shown in Fig. 12.2.

What should be noted in Fig. 12.2 is that "recycled" and "recyclable" have different meanings. The former refers to the recycled material formed via the regeneration of waste resources (e.g., recycled paper and secondary metal), and the latter refers to a product made of material that can be recycled after scrapping as waste resources being input into material production to form recycled material

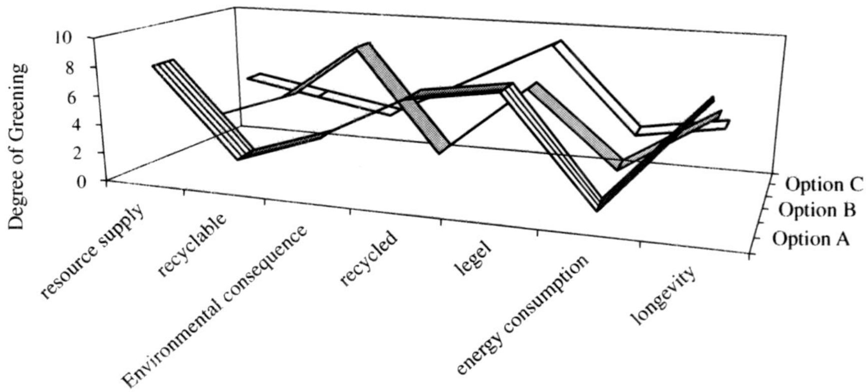

Fig. 12.2 Assessment of materials for its greening degree according to their environmental performance

(e.g., the majority of metal substances possess recyclable performance, and the fiber in paper will get shorter with the increase in regeneration times, and its recyclability will decrease). Recycled material can only be formed by the material possessing recyclability.

By means of the diagram in Fig. 12.2, we can understand the eco-environmental performance of a specific material. By comparing different substances on the seven aspects assessment diagrams, we can understand the ecological performance differences among different materials as the basis of material selection in product eco-design.

(III) Basic principles of material application

In practice, we often encounter the situation in which a specific material has advantages in one property but disadvantages in another; the properties of material in the application of eco-design must be evaluated for selection among different materials. Generally, the following basic principles should be considered:

(1) Recycled material should be given priority to reduce natural resource mining; e.g., recycled paper is used to produce presswork and secondary metal is used to make metal products. In recent years, cyclic regeneration has been achieved in an increasing number of varieties of metal substances, and the ratio of recycled material has increased gradually. With the implementation of a circular economy, it has increased 10–25% over 10 years. Table 12.1 displays the percentage of several metal substances in recycled materials.
(2) Scarce material should be only applied when necessary, and applying this type of material for low additional value should be avoided.
(3) Attempt to select and use materials with low energy consumption and low pollution, and conform to the state laws and regulations as much as possible. If toxic and harmful substances must be used in the production process, perform

Table 12.1 Percentage of several metal substances in recycled materials (Unit: %)

Substance name	Circulation rate	Substance name	Circulation rate
Al	45	Ni	44
Co	6	Steel	86
Cu	68	Sn	23
Pb	65	W	18

Data source Recycling nonferrous metals. Washington, DC: Bureau of Mines, 2006

on-site production as much as possible, and employ pollution prevention measures. Table 12.2 shows several substances on the list of toxic and harmful substance in our country.

(4) Attempt to select a material that is simple to recycle, reuse, or reproduce or is simple to be degraded after scrapping; for example, biodegradable garbage bags for kitchen garbage were accepted at the community level several years ago.
(5) Choose other substances to substitute for toxic and harmful or rare substances, namely material substitution, which is generally related to new resource development, new technique application, economic cost reduction, and other factors; e.g., in recent years, lithium batteries are gradually encouraged as a substitute for lead-acid batteries.
(6) In product or technique design, reduce the usage of various substances as much as possible. For example, for light and small products, new technical products are designed, or product marketing is replaced by marketing service. The strategy that resource efficiency and efficiency of labor are improved by improving additional value of material, product durability, reducing consumption of fossil fuel, and reducing other dissipative mediums to achieve economic growth is called dematerialization. It is described in the recommended references.

12.3.2 Product Packaging and Transportation

(I) Product packaging

Product packaging is an important measure implemented by the product manufacturer for product protection, convenient transportation, and marketing before the product is delivered to customers. After products are delivered to customers, it is the end of the service life of the packaging materials, resulting in packaging waste. According to statistics, packaging waste comprises approximately one-third of the urban solid wastes. Therefore, the eco-design of product packaging will be beneficial to the reduction of waste discharge in the product life cycle.

Product packaging is improved for the ecological environment, which is called green, pollution-free, or environmentally friendly packaging; it is the packaging conforming to environmental requirements and requests that commodity packaging

Table 12.2 Example of national hazardous wastes

Category of wastes	Industry sources	Waste code[a]	Hazardous wastes	Hazard feature[b]
HW01 medical waste	Health	831-001-01	Infectious waste	In
HW02 medical waste	Manufacture of raw material chemicals	271-002-02	Waste mother liquor and reactive waste produced in the production process of chemical synthetic raw material machine	T
HW03 wastes and drugs	Non-specific industry	900-002-03	Invalid, metamorphic and unqualified, eliminated and false medicine and drug (exclusive of HW01, HW02, 900-999-49) produced in the production, sales, and use procedures	T
HW04 pesticide waste	Pesticide manufacturing	263-009-04	Waste mother liquor produced in the pesticide manufacturing process and the cleaning effluent of reaction tank and container	T
HW05 wood preservative waste	Wood processing	201-002-05	Wastewater treatment sludge produced in the wood preservation process of creosote and the waste wood pieces stained with the preservative produced in the wood preservative treatment process	T
HW06 spent organic solvent and wastes containing the organic solvent	Non-specific industry	900-409-06	Wastewater treatment dross and sludge produced in the waste regeneration treatment process of 900-401-06 (exclusive of wastewater biochemical treatment sludge)	T
HW07 heat-treatment cyanide-containing wastes	Metal finishing and heat treatment processing	336-002-07	Quenching wastewater treatment sludge produced in the metal heat treatment process via cyanide	T
HW14 wastes of new chemicals	Non-specific industry	900-017-14	Chemical waste causing unclear impact on human beings or the environment produced in research, development, and teaching activities	T/C/I/R
HW16 wastes of photosensitive material	Printing	231-002-16	Developing (fixing) agent, film and waste developing paper produced in the process of printing developing via developing agent, etching-resist pattern developing, and letterpress printing	T

Source directory of the national hazardous wastes, 2016

[a]An 8-digit code is used to represent the hazardous waste. Digits 1–3 are the code of the industry generating the hazardous waste [determined in accordance with the *National Industries Classification* (GB/T 4754-2011)], digits 4–6 are the sequence code of hazardous waste, and digits 7–8 are the code of the hazardous waste categories

[b]Hazardous characteristics include corrosivity (C), toxicity (T), ignitability (I), reactivity (R), and infectivity (In)

causes no harm to the balance and to human health. Packaging methods can be selected based on priority during design. First, attempt to achieve no packaging, and second, attempt to minimize packaging. For required packaging, obey the following design principles:

(1) As a constituent part of the product itself, the packaging can be avoided; e.g., the packaging cup of ice cream is one part of the edible ice cream.
(2) The packaging is saved and simplified via improving old technology and using new techniques to avoid excessive packaging, e.g., to simplify the packaging of the traditional food moon cake.
(3) Reusable and recyclable materials are used; e.g., recycled plastic is used as packaging material.
(4) Attempt to use a single material, making waste recycling convenient.
(5) When various materials must be utilized, create a special packaging structure design to allow different materials to be separated easily, making it convenient for waste dismantling, classification, and recycling of different materials.

(II) Transportation

Transportation is the activity to achieve a location change of human beings and things. In the activity of a human society economy, transportation is throughout the entire life cycle process of the products. There is not only the transportation of final products but also the transportation of various materials, sources of energy, and wastes demanded for the product manufacturing process. Transportation is an important element to influence the environment.

To reduce the impact of the transportation process on the environment, the following aspects should be considered:

(1) Efficient and clean modes of transportation are selected to reduce the pollutant discharge.
(2) Effective measures are employed to prevent materials' scattering and leaking during the transportation process; e.g., when building wastes are conveyed by a construction vehicle, the vehicle is closed to contain the construction wastes.
(3) Packaging and shipping operations of toxic and harmful substance are standardized to ensure security.
(4) Standardized transport packaging is used.

12.3.3 Design of Other Elements

(I) Green process design

Green process is an important element to achieve green product manufacturing. Green process is also called the cleaner process as it is a technology capable of improving economic benefit while reducing environmental impact. Approaches to realize a green process include the following:

① To change the input method of raw materials and execute spot utilization, reuse the by-products and recovery products with practical value, and recycle various materials in a craft process. ② To change a manufacturing technique or technology, improve process control, and remold existing equipment, to minimize the consumption of raw materials, waste production, energy consumption, health and safety risk, and ecological damage. ③ To attempt to perform a corresponding environmental evaluation of the air, soil, water, and waste discharge into the natural environment, based on the relative scale of the environmental load, verify its impact on biodiversity, human health, and natural resources. A number of studies and practices show that the consumption of material and energy sources varies with a change in the product manufacturing process, so does the environmental impact. Green process planning attempts to plan and use the process program and process route with less consumption of material and energy sources, less wastes, and less environmental pollution based on the actual manufacturing system.

A green manufacturing process is related to the present technology, and there is currently no strict measurement level. Based on the characteristics of the manufacturing process, related domestic and overseas documents propose the following requirements. ① There should be no potential risk in the production environment and product use, causing no health threat to the operators, product users, or the environment. ② To reduce the usage of nonrenewable resources, attempt to employ various substituted substances and techniques. ③ The wastes produced in the production process should be recycled as much as possible, and the final wastes should be simple to process. ④ The process system should be simplified and optimized to improve system operating efficiency. ⑤ The process should be beneficial for reducing cost and improving economic and environmental benefits.

(II) Transformation from product marketing to service marketing

The essence of human demands is that human beings require a specific service instead of a product. People purchase products because a specific product has a specific service function. The product is the medium to carry the "service." For example, the purpose of buying cars or the specific-line tickets is to arrive at B from A; from the perspective of the enterprise, a car purchase belongs to product sales, and a ticket purchase is a service purchase.

If the enterprise can directly provide service to users instead of a product, it is likely that the number of transfers among different enterprises or individuals within the product life cycle decreases, or the serviceability of the product system under same environmental impact improves, to avoid or reduce unnecessary environmental impacts of a specific service. The form of directly selling a service to the users is called "service marketing," and it is an important method for eco-design specific to the element of product sales and for reducing the entire product system's environmental impact. The most basic characteristic of service marketing is that what is transferred in the economic activity is a product right instead of a product property right transferred via product marketing. There are many examples of common service marketing; e.g., cleaning service is conducted by the cleaning company in a specific location and documents are copied by a copy shop.

Usually, based on the relationship between customers and the service location, the service provided by the enterprise is divided into the following types. ① Customers go to the service location, e.g., hairdressing and clothes maintenance and cleaning. ② The enterprises go to the customer's location for service, e.g., house cleaning and maintenance of large electric appliances. ③ The enterprise and customer achieve service remotely, e.g., network distance education and security monitoring and control.

With the construction of large communal facilities, an increasing number of people are willing to purchase services instead of a product, so that more enterprises will sell services instead of products, e.g., public transport, house rental, and maintenance.

(III) Product use and disposal after use

For some products, the environmental impact generated during its use is the major portion of the entire product system's environmental impact. The environmental impact in the stage of product use can be not despised; improvement specific to environmental performance during product use is also an important content of an eco-design.

The following should be considered during the eco-design implementation of the stage of product use:

(1) Avoid designing a dissipative product; i.e., during use, all or some of the substances in the products will be discharged into the environment. The common dissipative products include printing ink, paint, washing products, cosmetic products, insecticides, and fertilizer.
(2) Energy-saving and cost-reducing design for product use. The energy consumption types in product use should be fully considered to choose clean energy sources. The quantity of energy consumption is considered for high-medium-low range design to adapt to different operating load demands. Product energy-saving operations and auto-power-off functions are designed to avoid energy waste.
(3) Extension of service life is designed for a product, and product durability is increased by improving the design of the product structure. A modular design is used to promote the convenient replacement and maintenance of parts and to extend product service life.

After product scrapping, the formed waste is not only the main source of environmental waste, but is an important source of waste recycling. For the convenience of reusing a portion of wastes, in product design, eco-design should be implemented for the product disposal after use. Consider whether each component constituting a product is easy to be dismantled after scrapping. Are the materials constituting a component simple to separate? Are these wastes simple to classify and recycle? Finally, for substances that cannot be regarded as secondary resources, how can harmless treatment and environmental emission be performed? For the convenience of product disposal after scrapping, post-scrapping disposal methods should be designed in eco-design; furthermore, these methods used in design should be integrated with the corresponding product component and material into a chip that acts as user manual for recycling of the end-of-life product. This chip is regarded as an important accessory and is sent to users with the products to guide them in performing post-scrapping disposal work.

Classroom Discussions and Assignments

Topic Discussion

The key technical measures for achieving environmental improvement are primarily discussed in this course from the design perspective. As one of the important methods for environmental improvement, in recent years, eco-design has gradually reached various fields. For example, many low carbon building designs were exhibited at the Shanghai World Expo in 2010, and energy saving utensils, electric cars, and other new products were frequently encouraged at the international fair held in Beijing in 2016. All of these indicate that eco-design is "blossoming and yielding fruit" to form many products suitable for the overall improvement of the environment and improving people's consumption and life. In this course, please choose a familiar product or service to discuss how to improve its environmental performance from the design perspective. List the research outline.

Homework and Presentation in Class

In the last chapter, we began to write papers. In this chapter, we will continue to exercise and improve writing skills.

The task of this chapter is to locate a published LCA research paper specific to the selected product or service for sharing, via PowerPoint, in the class by the group and to summarize it into a 1500- to 2000-word essay. To present the research status of a paper, include the topic selection process and the understanding and viewpoints of the paper. The grading scale is shown in Table 12.3.

Table 12.3 Grading of the group report in class

PowerPoint reporting quality in class (weight 0.5)			Essay writing (weight 0.5)			Total points
Process of topic selection (maximum 10 points)	Research basis (maximum 10 points)	Result and conclusion (maximum 10 points)	Topic selection and viewpoints (maximum 10 points)	Paper summary (maximum 10 points)	Expression and specification (maximum 10 points)	

Recommended and Referenced Literatures

1. Chunzheng, Lin, and Liu Yide. 2013. *Smart city and eco-design*, trans. Jia Liji, Peng Lin, and Liu Hailong. Beijing: China Architecture & Building Press.
2. S. Roaf, Fuentes M, and Thomas-Rees S. 2015. *Ecological architecture design guide*, trans. Wu Xiaojin. Beijing: Electronic Industry Press.
3. Yamamoto, Ryoichi. 2003. *Strategic environment management eco-design-models 100*, trans. Wang Tianming. Beijing: Chemical Industry Press.
4. Ayres, R.U. 2007. On the practical limits to substitution. *Ecological Economics* 61: 115–128.
5. Ayres, R.U., C.J.M. Jerpem, and B. van Den. 2005. A theory of economic growth with material/energy resources and dematerialization: Interaction of three growth mechanisms. *Ecological Economics* 55: 96–118.
6. Graedel, T. E., and B. R. Allenby. 2004. *Industrial ecology*, 2nd ed., trans. Shi Han. Beijing: Tsinghua University Press.
7. Ministry of Environmental Protection. 2016. *Interpretation to the directory of the national hazardous wastes* (2016 Version) [EB/OL]. (2016-06-21). http://www.mep.gov.cn/gkml/hbb/qt/201606/t20160621_354844.htm.
8. Ministry of Environmental Protection. *Directory of the national hazardous wastes* [EB/OL]. (2016-06-14) [2016-08-09]. http://www.mep.gov.cn/gkml/hbb/bl/201606/t20160621_354852.htm.
9. Zhong Wu, Lu. 2010. *Foundation of industrial ecology*. Beijing: Science Press.
10. Yamamoto, Ryoichi. 1994. *Environmental material*, trans. Wang Tianming. Beijing: Chemical Industry Press.
11. Dagen, Su, and Zhong Mingfeng. 2007. *Materials eco-design*. Beijing: Chemical Industry Press.
12. Changxiang, Yuan. 2009. *Practice guideline for the eco-design of energy-using products*. Beijing: China Standards Press.
13. Ling, Zheng. 2003. *Study on resource value accounting for circulating tax based on eco-design*. Beijing: Economic Science Press.
14. Graedel, T.E., and B.R. Allenby. 2003. *Industrial ecology*, 2nd ed. New Jersey: Prentice Hall.

13 Assessment of Environmentally Friendly Enterprises

13.1 Presentation of Core Issues

In the previous chapters, we have discussed a method to evaluate the environmental performance of the product (or service) provided by an enterprise, specifically, the product LCA method, and we have also discussed methods to avoid poor sourcing methods for the environmental performance of the product (or service), namely to perform eco-design for each element of the product or service and its entire life cycle. The eco-design implemented for all the activity (including production, sales) of the enterprise is important to achieve an environmentally friendly enterprise. However, the implementation of an integrated life cycle cannot ensure that the enterprise is environmentally friendly, and it must also place the completed eco-design into practice during each phase of the product life cycle, including both production and management of production, and consultation and material exchange with other enterprises, social organizations, administrative departments, or users. In general, an enterprise's behavior will have an effect on society, the economy, and the environment; this type of external effect will become an important benchmark to identify the operational behavior of the enterprise. Methods to perform an environmentally friendly assessment of an enterprise are the key topics in this chapter.

In this chapter, combined with establishing environmentally friendly enterprises in our country, we will discuss the relevant topics, such as, who should be responsible for the assessment, what should be assessed, and what procedure should be followed? Then, on this basis, we will discuss the advantages and disadvantages of existing practices in our country, and methods to improve the successive steps. In addition, an environmentally friendly enterprise is the initial stage of a sustainable development enterprise. In the long term, how does the enterprise evolve into a sustainable development enterprise by means of the assessment of an environmentally friendly enterprise? These topics are worth considering.

J. Mao et al., *Circular Economy and Sustainable Development Enterprises*,
https://doi.org/10.1007/978-981-10-8524-6_13

13.2 Why Do We Perform Assessments of Environmentally Friendly Enterprises?

In May 2003, the *Notice on Holding the State Environmentally Friendly Enterprise Creation Activity* was issued by the State Environmental Protection Administration [NO. 92 H.F. (2003)]. It is clearly indicated that the establishment of state environmentally friendly enterprises has a goal to "build a group of enterprise models with high technological content, good economic benefit, low resource consumption, less environmental pollution, and a 'win-win' environment and economy, so that the enterprise is encouraged to perform cleaner production, deepen industrial pollution prevention and to step onto the new industrialization road". From the aspect of the enterprise, the establishment of an environmentally friendly enterprise can reveal the environmental performance of the enterprise's behavior, ensure its degree of environmental friendliness, and resolve and review its safeguarding measures to perform eco-friendly construction activity, so that an industrial enterprise is encouraged to advance toward the new industrialization road.

An environmentally friendly enterprise is the core component of a circular economy, an important mechanism to stimulate environmental improvement of enterprise in our country, and is the highest affirmation of the enterprise to environmental performance. The implementation of the "state environmentally friendly enterprise" development activities has profound significance to promote industrial sustainable development in our country.

13.3 What Qualifications Should an Assessor Possess?

13.3.1 Professional Qualifications of an Assessor

An assessor must possess the general requirements of morality, culture, scientific research, and management, and simultaneously possess specialty literacy for enterprise assessment, including the following elements:

(1) Understand the role and function of an environmentally friendly enterprise in the sustainable development strategy. The establishment of an environmentally friendly enterprise is an important method to encourage the sustainable strategy. The assessor should identify the functions of the environmentally friendly enterprise in the sustainable development strategy clearly, which is beneficial for understanding the job objective and direction of the enterprise's ecological environment improvement. The assessor must evaluate the advantages and disadvantages between an enterprise's social economy and environmental improvement.

(2) Thoroughly understand the technical process of the enterprise behavior, including the various products (i.e., final or semi-finished) produced by the enterprise, and the production technology, process, equipment used in the

enterprise's manufacturing process and its operational state, the entire life cycle involved in the product, production technology, process, and equipment. This is the work object of an enterprise assessment, or the key elements that cause environmental impact would be overlooked.

(3) Thoroughly understand the standards of the ISO 14000 environmental management series. Strict business management is an important guarantee to achieve an environmentally friendly enterprise. The assessor should thoroughly understand the international environmental management standard, and based on the requirements of this standard, identify the insufficiency of the business management and assist the enterprise in improving its level of environmental management.

(4) Thoroughly understand the assessment method. The normative eco-friendly assessment method is the basis for implementing an eco-friendly assessment of the enterprise. It not only includes an assessment of the product provided by the enterprise, but includes an assessment of the production technology, process, and equipment and the enterprise's contribution to society and the economy. Therefore, an assessor is requested to not only thoroughly understand the international standard of product LCA but also to understand the socio-economic impact assessment method of the enterprise's behavior.

(5) Thoroughly understand the assessment procedure. The assessment of an environmentally friendly enterprise should be declared and evaluated in accordance with the country's procedural requirements for enterprise assessment, and the assessor should guide the enterprise to prepare based on the established sequence of the assessment procedures.

(6) Thoroughly understand the management rules and regulations of the country. All the activities of the enterprise should be in accordance with the national and regional laws and regulations. To perform an assessment of an environmentally friendly enterprise, the assessor should first understand the laws, regulations, and current systems of the country relevant to the enterprise's producing activity, to review and assist the enterprise in avoiding illegal activity.

(7) Other factors: The assessor should further focus on domestic and overseas development trends and assist enterprises by proposing comments and suggestions for future product upgrading, equipment upgrading, technology importing, and other activity.

13.3.2 Assessors in Our Country

In the *Implementation Plan (Trial) for the Establishment of a "National Environmentally Friendly Enterprise"* issued by the State Environmental Protection Administration in May 2003, it is stated that all industrial enterprises can apply to the provincial administrative department of environmental protection voluntarily, and after being reviewed and recommended by the provincial administrative department of environmental protection, they can be assessed, accepted, and released by the expert group for assessment as a national environmentally friendly enterprise and examined and approved by the State Environmental Protection Administration (SEPA). An enterprise that is completely in accordance with relevant requirements can be awarded the title of "national environmentally friendly enterprise." The administrator of the environmentally friendly enterprise is the provincial administrative department of environmental protection and the SEPA, and the assessor is the expert for the assessment of a national environmentally friendly enterprise.

In the initial stage of environmentally friendly enterprise assessment in our country, the CEC of the All-China Environment Federation (ACEF) is entrusted by the original SEPA to perform the organizational work of the environmentally friendly enterprise assessment. The ACEF is a national non-profit social organization charged by the Ministry of Environmental Protection. It is an organization that is formed voluntarily by people, enterprises, and public institutions enthusiastic for the environmental protection cause, approved by the State Council and registered by the Ministry of Civil Affairs in 2005 to promote full communication between government and enterprises. The CEC is responsible for receiving the applications of environmentally friendly enterprises, organizing the survey, and coordinating acceptance by the expert group.

In 2013, the *Notice on Initiating the Establishment of "China Alliance of Environment Friendly Enterprises"* (No. 13 Z.H.L.Z [2013]) was sponsored by ACEF to appeal to the enterprise that is in accordance with the requirement of an eco-friendly enterprise and is willing to become an environmentally friendly enterprise to establish the "China alliance of eco-friendly enterprises," so that an enterprise can be encouraged to actively conduct cleaner production, deepen industrial pollution prevention, and move toward sustainable development with cyclic, green, and low carbon development.

13.4 Method to Perform the Assessment of an Environmentally Friendly Enterprise

In addition to the *Notice on Holding the State Environmentally Friendly Enterprise Creation Activity* (No. 92 H.F. [2003]) issued by the State Environmental Protection Administration in May 2003, four supporting documents, *Implementation Plan (Trial) for the Establishment of a "National Environmentally Friendly Enterprise,"* *Index Explanation of the "National Environmentally Friendly Enterprise," Certification Range of an Environmental Label Product,* and *Application Form of the National Environmentally Friendly Enterprise* were issued. In the *Implementation Plan for the Establishment of a "National Environmentally Friendly Enterprise,"* it is clearly stated that the implementing object of the activity is all industrial enterprises that are an independent legal form of business. Additionally, the essential conditions of the enterprise, application, approval procedures, commending, review, and supervising process are stipulated. This plan becomes the basic method to guide the assessment of environmentally friendly enterprises in our country, along with the *Index Explanation of the "National Environmentally Friendly Enterprise."*

13.4.1 Formulating the Indicator System

A scientific assessment indicator system is an important basis to objectively assess and reflect the construction level of an environmentally friendly society. An environmentally friendly society is very complex and has numerous levels; furthermore, there are interactions among subsystems and mutual input/output. Therefore, it needs to select a sensitive, measurable, and rich-connotation dominant index from numerous indicators as the assessment factor. The indicator is selected via analysis of the connotation, structure, and development methods of the environmentally friendly enterprise combined with the general principles for building an indicator system, to build a complete assessment indicator system. A scientific environmentally friendly assessment system should meet the following basic principles:

(1) Characteristic—Currently, there are many social construction-related assessment indicator systems, e.g., the sustainable development and ecological city indicator systems. Compared with these, an environmentally friendly enterprise focuses on the friendliness degree of the producing activity to the natural environment and the friendly response of the natural environment to the enterprise's production; therefore, its assessment indicator system must possess its own characteristics.
(2) Science—An indicator system should comprehensively reflect the various aspects of an environmentally friendly enterprise in accordance with its connotation and objectives, and concurrently, the composition and construction approaches of the environmentally friendly enterprise should be closely connected to reflect the real situation of the environmentally friendly enterprise construction.
(3) Comparability—On the basis of considering the differences in economic development level, social culture, actual situation of resources, and the environment among different types of industry and different areas of an enterprise, a common aggregative indicator should be selected in the indicator system to ensure the regional comparability of the evaluation results.
(4) Operability—A major indicator system capable of reflecting the features of an environmentally friendly enterprise must be chosen, and subordinate indicators closely related to the major indicators must be evaluated, to allow the indicator system to be concise, clear, and simple to apply. Additionally, the indicators should possess measurability and comparability. There are certain quantitative measures corresponding to the qualitative index, and the quantitative index should be calculated directly or indirectly via the data released by the state statistical department.

(5) Combined systematic and hierarchy—An environmentally friendly enterprise is a subsystem of an environmentally friendly society. The assessment system should not only reflect the characteristics of the enterprises' subsystems and other subsystems in society but should also reflect the overall development situation of this multisystem and multilevel dynamic composite system, i.e., the environmentally friendly society, to organically relate the assessment objective and the indicator to allow the enterprise and other social systems to constitute an integrated and well-arranged system.

13.4.2 Indicator System for an Environmentally Friendly Enterprise

Strict indicator requirements are established for a "national environmentally friendly enterprise" by the State Environmental Protection Administration from three aspects: environment, management, and product. They include reaching and obeying the requirement of each relevant environmental protection law, norm, and standard; conscientiously implementing cleaner production and using advanced clean manufacturing techniques; and establishing a perfect environment management system. Product safety, health, and quality should be in accordance with the relevant standard requirements, and an enterprise is not allowed to use the substances prohibited by relevant laws, regulations, standards, and international conventions signed by our country. A product within the scope of an environmental label product should be in accordance with the requirements of the approval standards for an environmental label product. These requirements are specifically embodied in 20 indicators, as shown in Table 13.1, and they are the major basis to guide and establish a national environmentally friendly enterprise.

13.4.3 Examples of an Indicator System

Accurately understanding the meanings of each indicator in the environmentally friendly assessment indicator system is an essential condition for the enterprise's environmentally friendly improvement and is also the premise to create an environmentally friendly enterprise. At the same time, the *Notice on Holding the State Environmentally Friendly Enterprise Creation Activity* was issued by the State Environmental Protection Administration, and the *Index Explanation of the "National Environmentally Friendly Enterprise"* was issued. Several indicators in Table 13.1 are selected for explanation:

Table 13.1 Assessment indicator system for China's environmentally friendly enterprises

Indicator categories	Assessment indicators
Environmental indicators	Target rate of the enterprise's total emission pollutants
	Comprehensive energy consumption of an enterprise's unit product
	Water consumption of a unit product
	Discharge of major pollutants of a unit's industrial output
	Comprehensive utilization efficiency of the enterprise's wastes
	Level of the enterprise's environmental management certification
Management indicators	Enterprise's level of implementing cleaner production
	Implementation rate of the "environmental impact assessment" and "three simultaneities" systems of the enterprise's new, transformed, and expanded project
	Operating rate of the enterprise's environmental protection facilities
	Disposal rate of the enterprise's solid and hazardous wastes
	Green coverage ratio within the plant area
	Standardized degree of the enterprise's sewage drain exit
	Validity of the enterprise's dumping right
	Payment status of the enterprise's sewage charge
	Enterprise's "good" environmental performance
	Standardization level of the enterprise's environmental management
	Satisfaction rate of the surrounding residents and enterprise's staff regarding the environment protection work
	Enterprise voluntarily continues to reduce pollutant emissions
Product indicators	A prohibited substance is contained in the product and its production process
	Level of product safety, health, and quality

(I) Examples of environmental indicators

(1) Target rate of the enterprise's total emission pollutants—This is the total amount of the exhaust gas, waste water, noise, solid waste, and radioactive waste discharged by the enterprise in accordance with the national or regional emission standards; it reaches the corresponding control standards, and it is represented as a percentage. The country's target rate for this indicator for a national environmentally friendly enterprise is 100%.
Please note that the up-to-standard of the total quantity mentioned indicates that the "quantity" and "quality" of the pollutant should be in accordance with the requirements of the environmental control management. The "total quantity control" estimates the maximum discharge of pollutants allowed to achieve the requirements of the regional environmental objective. Then, this quantity is allocated to each source of pollution. This is the method the country uses to control the total quantity of pollutants. In contrast, the method of controlling

environmental quality by controlling the concentration of the pollutants discharged from the enterprises is called the management method of "concentration control."

(2) Comprehensive energy consumption of an enterprise's unit product—This refers to the comprehensive energy consumption conversion based on standard coal when 1 t of product is produced by the enterprise (g of standard coal). Unit-product comprehensive energy consumption of a national environmentally friendly enterprise is requested to be leading in its industry. Put differently, the unit-product comprehensive energy consumption (g of standard coal/t) is not more than the level of energy consumed in the same industry. For example, the comprehensive energy consumption for caustic soda production by a specific enterprise (ion-exchange membrane process 30%) is 294.52 kg standard coal/t, and the standard coal consumption for the thermal power generation of a power plant is 292.57 g standard coal/ (kWh), which are lower than the national average of 3.08 kg standard oil/t and 20.24 g/(kWh), respectively.

(3) Comprehensive utilization efficiency of an enterprise's wastes. This refers to the amount of the three wastes that are extracted from the three wastes (waste water (liquor), exhaust gas, and waste residue)) produced during the production process or the portion transformed into recyclable resources, a source of energy, and other materials via recovery, processing, circulation, and exchange, represented as follows:

$$\text{Comprehensive utilization efficiency of an enterprise's wastes} = \frac{\text{amount of waste comprehensively utilized by the enterprise within the reporting period}}{\text{enterprise's waste output within the reporting period}} \times 100\% \quad (13.1)$$

The comprehensive utilization efficiency of wastes in a national environmentally friendly enterprise must be greater than the level of wastes comprehensively utilized in the same industry.

For example, a circular economy is greatly developed by China's Hunan Valin Hengyang Steel Tube (Group) Co., Ltd. in its Valin Hengyang steel ironmaking project, and comprehensive utilization is implemented for the various materials generated during the ironmaking process; the comprehensive utilization efficiency of solid wastes has reached 98%.

(4) Level of the enterprise's environmental management certification—This indicates that the enterprise has obtained an ISO 14001 environmental management certification or established a perfect environmental management system based on the standards of the ISO 14001 environmental management system. A national environmentally friendly enterprise is requested by the country to issue a valid certificate of ISO 14001 environmental management system certification or to provide all relevant materials concerning the perfectly established environmental management system.

(II) Examples of management-type indicators

(1) Enterprise's level of implementing cleaner production—This refers to the degree that the enterprise conscientiously implements cleaner production and employs an advanced cleaner production process, including when the enterprise's existing production capacity, process, and product are beyond the scope of the eliminated directory within the time limit explicitly stipulated by the country (as shown in the later example of an eliminated directory) and when the qualified cleaner production review institution and staff are requested to perform a cleaner production review to implement the waste-free, low-waste, medium-waste, and partial high-waste plans proposed in the cleaner production review, resulting in good environmental and economic benefits. The national environmentally friendly enterprise is requested to provide an audit report of cleaner production and a final report of the implementation of the cleaner production plan.

(2) Implementation rate of the environmental impact assessment and the three simultaneities systems of the enterprise's new, transformed, and expanded project—This refers to the executive level of the enterprise's new, transformed, and expanded project environmental impact assessment and three simultaneities. The new, transformed, and expanded project is required to be in 100% accordance with the national regulations on the environmental protection of a construction project; to not avoid the state construction projects' environmental impact assessment, three simultaneities and other systems within the previous three years; and to receive the acceptance of the environmental protection department for project approval.

    What must be stated here is that the "three simultaneities" system refers to the regulation of Article 26 in our country's Environmental Protection Law, which states "the measures for preventing and controlling pollution in the construction project must be designed, built and incorporated simultaneously with the main work." See details in Chap. 14, integrated management of an enterprise.

(3) Standardized degree of the enterprise's sewage drain exit—This indicates that a standardization treatment is used to establish the enterprise's sewage drain according to the environmental requirement of the country. The main sewage drain with a significant pollution discharge load is provided with waste water and exhaust gas pollutant online monitoring devices. Currently, the major monitoring indicators include discharge amounts of wastewater, COD, ammonia nitrogen, pH, emission of waste gas, SO_2, smoke, and ash. A national environmentally friendly enterprise is requested by the country to arrange standard signboards at the sewage drain exit, to install the pollutant online monitoring devices, and to incorporate these activities into their normal

operation. For example, during the establishment of an environmentally friendly enterprise, a specific coal mine plant performs the standardization treatment on all the sewage drains of the enterprise and posts the standard signboards, as shown in Fig. 13.1.

(a) Signboard for noise emission source

(b) Signboard for general solid waste source

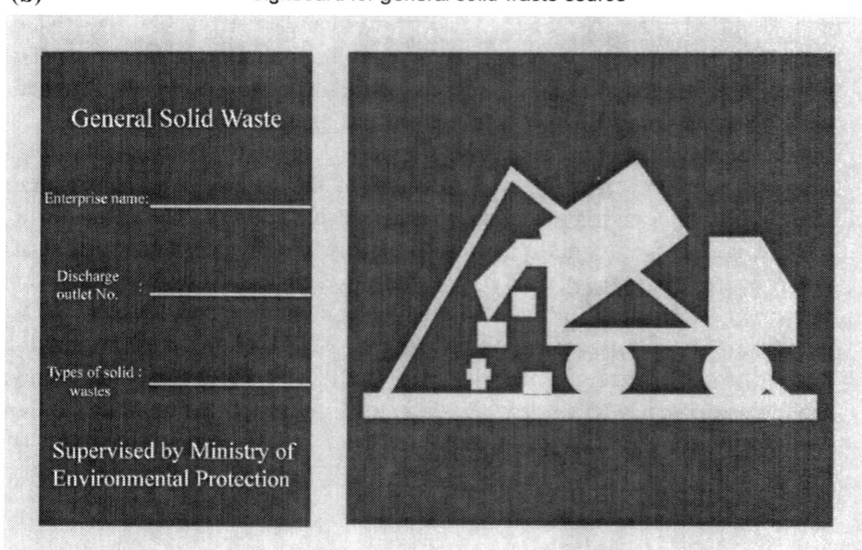

Fig. 13.1 Normative signboard for a mine sewage draining exit

(4) Validity of the enterprise's dumping right—This indicates that according to the requirements of national environmental protection regulations, the enterprise has applied and registered for sewage disposal, possesses a pollutant discharging license, and executed the system of a pollutant discharging license within the corresponding area. A national environmentally friendly enterprise is requested to show the valid relevant materials of the application and registration for pollution discharge, the original sewage discharge permission, and their operations manual related to pollution discharge based on the license.

(5) Payment status of the enterprise's sewage charge—According to legal provisions of the state, an enterprise pays the full amount of the sewage charge to the environmental protection department on time. Types of sewage charges include sewage water discharge fees, charges for disposing waste gas, sewage charges for solid waste and hazardous waste, and an over standard discharge fee for noise. A national environmentally friendly enterprise is requested by the country to show its sewage charge payment was paid in full by its due date.

(6) Satisfaction rate of the surrounding residents and enterprise's staff regarding the environment protection work—This refers to the satisfaction rate of the surrounding residents and enterprise's staff regarding the production and living environment; usually, such an indicator is obtained by the examination team via spot field investigations. The number of people investigated should be not less than 200. It is stipulated by the country that the satisfaction rate of the surrounding residents and enterprise's staff regarding environment protection work must be a minimum of 90%. It is important to note that public participation is involved in the execution of this indicator, which refers to the purposeful social action of the social group, social organization, unit, or individual as the main body within the scope of rights and obligations.

(III) Examples of product-type indicators

(1) A prohibited substance is contained in the product and its production process. The product of a national environmentally friendly enterprise is requested by the country to be in accordance with the following two requirements: It is not allowed to contain or use substances prohibited by national laws, regulations, and standards in the product and its production process, and it is not allowed to contain or use a substance prohibited by the international convention signed by our country in the product and its production process.

(2) Level of product safety, health, and quality. The product produced by a national environmentally friendly enterprise is requested by the country to be in accordance with the requirements of the country, industry, or enterprise in the areas of product safety, health, and quality. Additionally, the product listed within the scope of the national environmental label certification (as shown in Table 13.2) should be verified according to the requirements of the environmental label product certification standards; a product that obtains the environmental mark does not need to be assessed.

Table 13.2 Examples of the certification range of environmental label products

Product name[a]	Product range[a]	Standard number[b]	Standard name[b]
Recycled paper products	Recycled paper made of waste paper, decorations, and building materials	HJ/T 205—2005	Technical requirements for environmental label product Recycled paper products
Hg-Cd-Pb-free rechargeable battery	Various types of rechargeable batteries	HJ/T 238—2006	Technical requirements for environmental label product Rechargeable batteries
Detergent	Fabric detergent, tableware detergent, and industrial detergent (water-based type)	HJBZ 8—1999	Detergent
Mercury-free dry batteries	Various types of dry batteries	HJ/T 239—2006	Technical requirements for environmental label product Dry batteries
Packaging products	To substitute for the non-degradable plastic film/bag packaging product and paper–plastic composite packaging products	HJ/T 209—2005	Technical requirements for environmental label product Packaging product
Soft drinks	Alcohol-free beverage	HJ/T 210—2005	Environmental label product standard Soft drinks
Children's toys	Toys designed for children 14 years or younger or obviously used by them	HJBZ 16—1996	Children's toys
Health insect aerosol	Low-toxicity, efficient, and safe health insect aerosol	HJBZ 20—1997	Healthy insect aerosol
CFC-free refrigeration plant for industry and commerce	Various types of refrigeration plants except for domestic refrigerating appliances, and the refrigeration plant for vehicles and ships	HJ/T 235—2006	Technical requirements for environmental label product Refrigeration plant for industry and commerce
Household microwave ovens	Household microwave oven with not more than 250 V nominal voltage	HJ/T 221—2005	Technical requirements for environmental label product Household microwave ovens

(continued)

Table 13.2 (continued)

Product name[a]	Product range[a]	Standard number[b]	Standard name[b]
Microcomputer and display	Computer host and display	HBC 15—2002	Microcomputer and display
Halon-substituting fire extinguisher	ABC dry powder extinguisher, aqueous film forming foam (AFFF) concentrate, CO_2 extinguisher, water fire extinguisher, and fire extinguishers with inert gas, HFC, HCFC, and other fire extinguishing agents	HJ/T 208—2005	Technical requirements for environmental label product Fire extinguisher
Adhesive	Adhesive for interior and exterior for architectural decoration, wood processing adhesive, water-based paper–plastic composite film adhesive, and adhesive for shoemaking and carpet making	HJ/T 220—2005	Technical requirements for environmental label product Adhesive
Building plastic pipe	Plastic pipe and plastic–metal composite tube and pipe fitting substituting cast iron pipe and galvanized tube	HJ/T 226—2005	Technical requirements for environmental label product Building plastic pipe
CFCs-free foamed plastic	Expanded plastic products except for disposable foaming tableware	HJ/T 233—2006	Technical requirements for environmental label product Foamed plastic
Aerosol	CFCs-free aerosol products, containing no insect aerosol	HJ/T 222—2005	Technical requirements for environmental label product Aerosol
Disposable tableware	Disposable tableware made of biodegradable, photo-biodegradable, and easily recovered materials	HJ/T 202—2005	Technical requirements for environmental label product Disposable tableware
Water-based paint	The paint with waster as a solvent or dispersion medium and related products	HJ/T 201—2005	Technical requirements for environmental label product Water-based paint
Wallpaper	Wallpaper with paper as its base material	HBC 23—2004	Wallpaper
Sanitary ware	Wash basin, pedestal pan, bidet, washing tank, squatting pan, bathtub, and other sanitary wares	HJ/T 296—2006	Technical requirements for environmental label product Sanitary ware

(continued)

Table 13.2 (continued)

Product name[a]	Product range[a]	Standard number[b]	Standard name[b]
Low-pollution motorcycle	Two- and three-wheeled motor vehicle with more than 50 ml of engine displacement and 50 km/h or greater design speed	HJBZ 38—1999	Low-pollution motorcycle

[a]State Environmental Protection Administration. Certification Range of Environmental Label Product, Notice for Holding the State Environment Friendly Enterprise Creating Activity [No. 92 H.F. (2003)]. 2003-05-23. http://www.mep.gov.cn/gkml/zj/wj/200910/t20091022_172194.htm

[b]Certification Range of China Environmental Label Product (55 types), 2004-10-10. http://www.cepf.org.cn/regulations/200907/t20090723_156589.htm

Currently, the product environmental label certification activities in our country should be controlled by the environmental certification center of the Ministry of Environmental Protection.

For the indicators listed in Table 13.1 but not explained, relevant or recommended documents may be downloaded from the Internet. It is further expressly stipulated in the document that the majority of the data or information in the above indicators must be provided by the enterprise via statistics, checked by the provincial environmental protection administrative authority, accepted and confirmed by the experts of the assessment group.

13.4.4 Examples in the Elimination Directory

As previously mentioned, for the establishment of an environmentally friendly enterprise, the enterprise is requested to not use the substances prohibited by the relevant laws, regulations, standards, and international convention signed by our country. The product within the scope of an environmental label product should be in accordance with the requirements of the approval standards for an environmental label product. In recent years, to prevent and control environmental pollution and to promote the upgrading of manufacturing techniques, equipment, and products, regulations have been successively issued by the State Economic and Trade Commission, environmental protection agency, State Administration of Work Safety, and other departments to eliminate lagging production equipment and products. Several cases are provided below to assist readers to be aware of additional relevant situations.

(I) Examples in the elimination directory of production capacity, processes, and products

Since 1999, the State Economic and Trade Commission has successively released three versions of the *Elimination Directory of Lagging Production Capacity,*

Processes, and Products to eliminate the lagging production capacity, processes, and products that violate national laws and regulations, have lagging modes of production and inferior quality, cause serious environmental pollution, and have high consumption of raw material and energy. The third version of the elimination directory involves 15 industries and 120 elements, such as fire control, chemical engineering, metallurgy, gold, building materials, press and publications, light industry, spinning, cotton processing, mechanical engineering, electric power, railways, automobiles, medicine, and hygiene. Several examples are shown in Table 13.3.

(II) Examples in the elimination directory of high energy consumption and lagging electromechanical devices (products)

To further advance energy conservation and emission reduction in industry and to accelerate the elimination of the lagging production capacity and high energy consumption equipment, the Ministry of Industry and Information Technology issued the Elimination Directory (1st version) of High Energy Consumption and Lagging Electromechanical Device (Product) (No. 67 G.J. [2009]). It involves nine categories and 272 types of equipment (products), including, motors, electric welding machines, voltage regulators, boilers, and fans. Each production and use department are requested to promptly eliminate the equipment (product) listed in

Table 13.3 Examples in the elimination directory of the lagging production capacity, processes, and products

Type of elimination	Name of elimination
Lagging production capacity	Chemical pulping papermaking production device with less than 10,000 t annual production capacity
	Diagonal tire whose annual production capacity is less than or equal to 500,000 pcs. or the tire with natural cotton cord fabric as its framework
	Soda production line with less than 100 bottles of production capacity/min
Lagging manufacturing techniques and equipment	Small amalgamation roller gold extracting process
	T100, T100A bulldozer
	Cyanide-containing electroplating (2003)
	Manual capsule filling
Lagging products	Carburetor-type and 5-seat passenger cars (refer to production and sales)
	PY5 digital thermometer
	Toilet with a greater than 9 L one-time flush volume
	Biological and blood products with ordinary natural rubber bung as packaging

Source State Economic and Trade Commission. Elimination Directory of Lagging Production Capacity, Process and Product (the 3rd batch) (No. 32 SETC order). 2002-06-02. National Energy Administration 2011-08-18. http://www.nea.gov.cn/2011-08/18/c_131057676.htm

Table 13.4 Examples of eliminated products with high energy consumption and lagging electromechanical devices

No.	Name of the eliminated lagging technical equipment	Cause of elimination
1	JD deep-well pump motor	Old structure, low efficiency, and low locked-rotor torque
2	Ac arc welder BX1—135, BX2—500	The Soviet Union-imitated old products from the 1950s, large volume and heavy, uses more consumable items and has poor performance
3	Distribution transformer SL7-30/10 ~ SL7-1600/10, S7-30/10 ~ S7-1600/10	Large consumption of raw materials, high no-load loss, high load loss, and low operational reliability
4	Shaking grate steam boiler KZZ2-13	Old shaking grate boiler, low thermal efficiency, and serious pollution
5	SD50 series channel axial flow fan	Low efficiency
6	Small submersible electric pump QY-7	Old structure, low electric efficiency, and low water pump efficiency
7	B-1.3/15 air compressor	Old structure, lagging performance, high energy consumption, and low efficiency
8	Refrigerator 4AJ-15	Old product mix, large volume, and lagging performance index
9	481 Diesel engine	This is the Adam Fergusson-imitated product from the 1950s; no major improvement has been made after being in production for more than 20 years. High fuel consumption rate compared with the factory standard

Source Ministry of Industry and Information. Elimination Directory (1st Batch) of the High energy Consumption and Lagging Electromechanical Device (Product) Announcement. No. 67 G. J. [2009]. 2009-12-04
http://www.miit.gov.cn/n1146295/n1146592/n3917132/n4061768/n4061782/n4061783/n4061785/c4169975/content.html

this Directory, the production unit should stop production, and the use unit should replace the eliminated equipment with energy-efficient equipment (product) as soon as possible. Each level of the energy-saving supervising authority should strengthen the supervision and inspection of the production suspension and elimination status of the equipment (product) listed in this directory. Examples of several eliminated products with high energy consumption electromechanical devices are shown in Table 13.4.

(III) Examples in the elimination directory of safety technical equipment

For safety technical equipment, to accelerate the elimination of lagging safety technical equipment, to promote the level of the enterprise's safety assurance, and to enhance the precaution and restraint of major and extraordinarily serious accidents, in 2015, the state safety and supervision bureau issued the Elimination Directory of Lagging Safety Technical Equipment (the 1st version in 2015)

(No. 75 A.J.Z.K.J. [2015] announcement). It eliminates 30 types of equipment in five major fields, such as coal mines, dangerous chemical and fireworks industry, metallurgy, occupational health field, and emergency rescue. Examples of several types of eliminated technical equipment are shown in Table 13.5.

Additional strict local elimination directories are issued in various places, e.g., Shanghai and Beijing. They are an important basis to assess an enterprise's environmentally friendly level.

13.4.5 Assessment Requirement

For enterprises' assessments, an organization declaring to be an environmentally friendly enterprise should meet the basic assessment requirements that follow:

(1) Obey environmental protection laws and regulations in accordance with the national policy. Most importantly, an enterprise must obey the environmental protection laws and regulations, and its own construction development should be consistent with the national and regional industrial policy and environmental protection policy, urban master planning and protection planning.
(2) Ensure that the environmental protection facilities are steadily running and pollutant discharge becomes stable and reaches the requirements of total quantity control and reduction. The relevant assessment indicators and their corresponding examples are described in sections II and III of this chapter, respectively. It is emphasized that the equipment operation should be stable and the pollution discharge must reach national standards.
(3) Achieve low consumption of resource energy and significant efficiency of cleaner production. The enterprise must actively perform the review of cleaner production and ensure that the implementation of the cleaner production measures and plans can achieve remarkable results and can reach the basic requirements of cleaner production-advanced enterprises.
(4) Incorporate high technological content in manufacturing techniques and equipment, friendly products, and the environment. The enterprise produces technical equipment that reaches or is above the domestic level by using cleaner production and eco-friendly processes and technology and develops and promotes eco-friendly techniques and products.
(5) The constructive process of the enterprise should be practical and valid to achieve a "win-win" situation for the economy and environment. Put differently, an enterprise must establish a work program and actively develop a circular economy to achieve significant results.
(6) Present the enterprise's environmental information to the public to achieve interaction. The enterprise must regularly disclose its environmental information to allow staff and surrounding residents understand and be satisfied with the environmental protection work.
(7) Actively assume the social responsibility of environmental protection. An enterprise is requested to actively promote relevant environmental protection,

Table 13.5 Examples of eliminated safety technical equipment

No.	Name of the eliminated lagging techniques and equipment	Cause of elimination	The recommended type of elimination	The recommended scope of limiting value	Name of the replacement technical equipment
1	Oil immersed transformer underground coal mine, oil switch, and other oil immersed electrical equipment	Low insulation level and breaking capacity, poor reliability, high operating cost, significant maintenance required, burn risk of the electric insulating oil, and the oil immersed equipment in the electromechanical chamber have been eliminated	Prohibition		Dry-type transformer and vacuum (or air) breaker
2	S7 power transformer	Poor safety reliability, significant material, and energy consumption	Limiting value	New product is not used	S9 and above power transformer
3	Synthesis ammonia-semi-water gas ammonia liquid-phase desulphurization process	There is no supporting sulfur recovery device, complex process control, significantly dangerous, harmful factors and unforeseeable dangers, low automation control level, poor security, likely to leak, poisoning, explosion, fire, and other work safety accident	Prohibition		Tannin extract wet desulfurization process for the sulfur recovery device
4	Hand-mixed explosive from fireworks production	The explosive has direct contact with people, there is large explosive quantity on site and it is very inclined to cause combustion and explosion, resulting in casualties	Prohibition		Mechanized mixed explosive
5	Well-type heat treatment electric furnace for heat treatment	During the electric furnace conversion process from "solid solution-quenching-tempering," products must be removed and placed into another furnace via a traveling crane by person with a poor safety record	Prohibition		Heat treatment process, continuous heat treating furnace of fuel gas
6	Spray painting process for wooden furniture manufacturing without independent spraying shop	The chemical poison produced in the painting process easily harms the personnel performing other processes	Prohibition		Independent spraying shop is arranged

(continued)

Table 13.5 (continued)

No.	Name of the eliminated lagging techniques and equipment	Cause of elimination	The recommended type of elimination	The recommended scope of limiting value	Name of the replacement technical equipment
7	Negative-pressure oxygen respirator	During use, the pressure in the entire respirator system represents a positive–negative alternation. For expiration, the pressure of the system is higher than the barometric pressure, while for inspiration, the pressure of the system is lower than the barometric pressure. If the oral–nasal becomes flexible or falls, the personnel are very easily damaged by the harmful gas, it has low safety	Prohibition		Positive-pressure oxygen respirator

Source The State Safety and Supervision Bureau. Elimination Directory of the Lagging Safety Technical Equipment (the 1st batch in 2015). Notice of the State Safety and Supervision Bureau on Printing and Distributing the Elimination Directory of the Lagging Safety Technical Equipment (the 1st batch in 2015) (No. 75 A.J.Z.K.J. [2015]). 2015-07-10
http://www.chinasafety.gov.cn/newpage/Contents/Channel_5330/2015/0717/254848/content_254848.htm

advocate production modes and lifestyles beneficial for resource saving and environmental protection, actively promote the improvement of regional environment quality, and participate in the public welfare works relevant to environmental protection.

(8) Adhere to the continuous improvement of resources, and constantly improve environmental performance. An enterprise is requested to establish and perfect a valid cleaner production continuous implementation system, adhere to the constant improvement of resources, and continuously reduce pollutant emission.

13.5 Method to Perform the Application and Review

13.5.1 Basic Conditions of the Application

An application from an environmentally friendly enterprise should be voluntary, and the enterprise should be in accordance with the following basic conditions: ① The enterprise's production and management conform to requirements of the national laws and regulations; ② the product produced by the enterprise is in accordance with the national relevant requirements of quality, security, and environmentally friendly; and ③ in the previous three years, there has been no punishment from the environmental protection department and no complaints of an environmental problem from the surrounding residents and social organization. These three basic conditions are necessary for the enterprise to declare itself to be an environmentally friendly enterprise; therefore, ensure that the enterprise completely conforms to the above conditions via self-examination before submitting the application.

Generally, it is not difficult for common enterprises to achieve the three basic conditions above, but to establish a national environmentally friendly enterprise, significant effort is required and the enterprise must understand its importance. Furthermore, achieving certain environmental performance by implementing some environmentally friendly measures internationally lays the foundation for establishing the environmentally friendly enterprise within a wider range. Additionally, the enterprise must establish the self-assessment standard of the national environmentally friendly enterprise and other series of indicators specific to the enterprise based on the indicator system established by the country and prepare a series of implementation measures. In the implementation process, environmental performance is assessed by a professional assessment panel.

For example, in national environmentally friendly enterprises, the model for the iron and steel industry is the Jinan Iron and Steel Co. Ltd., who first began the new path to industrialization proposed by the Sixteenth CPC National Congress before applying to be named an environmentally friendly enterprise; the enterprise provides environmental protection and other publicity and education to the staff and adjacent community through a series of activities. In May 2004, the enterprise held a startup meeting to establish an environmentally friendly enterprise and then established leading and working groups to prepare relevant remediation action, self-examination, and self-assessment standards to bring to fruition a national environmentally friendly enterprise. These measures form the foundation for the application of an environmentally friendly enterprise.

13.5.2 Application Procedure

Within the application to become an environmentally friendly enterprise, the enterprise must propose a written application to the All-China Environment Federation's CEC and submit the application for the assessment (in triplicate), the technical report, and the self-examination report.

(I) Application form

The application form for an environmentally friendly enterprise includes four sections: the description of the enterprise, environmental and management indicators, review from provincial environmental protection department, and audit opinion from the industry association. The description of the enterprise includes the type of enterprise, enterprise-involved environmental impact, and environmental regulations of the obeying business. The description of enterprise environment and management involves mainly the creation of environmentally friendly enterprise, it includes their pollutant discharge (as shown in Table 13.6), progress of work for environmental protection (as shown in Table 13.7), the current environmental achievement of the enterprise, the enterprise's commitment to future environmental performance (as shown in Table 13.8), and their participation with the public (as shown in Table 13.9). Here, each enterprise applicant is requested to prove its current environmental achievement and to make a commitment to future improvement. Additionally, each enterprise is required to describe its commitment for environmental performance improvement specific to at least two major pollutants; the enterprise can also provide supplementary instruction for other environmental performance in the future.

Table 13.6 An enterprise's current pollutant discharge

Name of pollutants	Actual emission concentration	National standard	Total amount of discharge (t/a)	Index value of the enterprise's total quantity control (t/a)

Table 13.7 Work progress of an enterprise's environmental protection

I. Records of regulations abidance						
1. The enterprise continuously abides by the rule of law	Has the enterprise continuously abided by the rule of law for 3 years?		Yes	If there was illegal behavior in the enterprise, when did it occur?		
			No			
2. Has there been a serious environmental pollution accident within 3 years?	Yes	If yes, when did it occur?	3. Has there been a repeated environmental petition letter event within 3 years?	Yes	If yes, when did it occur?	
	No			No		
II. Pollution control and cle aner production						
4. Water consumption per unit product			What is the level?	International advanced level		
				Domestic advanced level		
				Domestic general level		
5. Comprehensive energy consumption per unit product			What is the level?	International advanced level		
				Domestic advanced level		
				Domestic general level		
6. Pollutants per unit industrial output			What is the level?	International advanced level		
				Domestic advanced level		
				Domestic general level		

(continued)

Table 13.7 (continued)

I. Records of regulations abidance			
7. Comprehensive utilization efficiency of wastes		What is the level?	International advanced level
			Domestic advanced level
			Domestic general level
8. Disposal rate of industrial solid waste		9. Safe disposal rate of hazardous waste	
10. Success ratio of the environmentally protective facility		11. Performance of the environmentally protective facility	
12. Is a major sewage drain exit provided with the pollutant online monitoring facility?		13. Normal operation rate of monitoring facility	
14. Has the review of cleaner production been approved?	Yes	Date of approval Certificate number	If being reviewed, the estimated time of completion
	No		
III. Environmental management			
15. Implementation rate of the environmental impact assessment system of the new, transformed, and expanded project		16. Implementation rate of the three simultaneities system of the new, transformed, and expanded project	
17. Are the pollution discharge application and registration implemented?		18. Is there sewage discharge permission?	
19. Is the sewage charge paid according to regulations?		Amount of sewage charge paid by the enterprise annually	
20. Is there a special environmental management organization in the enterprise?	Yes	Name of the organization	
	No		

(continued)

Table 13.7 (continued)

I. Records of regulations abidance			
21. Is there a special environmental management system in the enterprise?	Yes	22. Is there a complete environmental protection file?	Yes
	No		No
23. Does the enterprise pass the ISO14000 certification?	Yes	Date of approval Certificate number	If being audited, the estimated time of completion:
	No		
24. Factory district greening and beautifying			Greening rate:
25. Satisfaction of the surrounding residents to the enterprise's environment protection work	Quite satisfied		
	Basically satisfied		
	Unsatisfied		
26. Satisfaction of the staff to the enterprise's environment protection work	Quite satisfied		
	Basically satisfied		
	Unsatisfied		
IV. Product indicator			
27. Do the major products of the enterprise pass the environmental label certification?	Yes		
	No		
If any, please list the product name and date of approval			

Table 13.8 Current environmental achievement of the enterprise and its commitment to future performance

Major pollutant		
Current level	Absolute value (quantity/unit)	
	Output of unit product (quantity/unit)	
Future (in three years) level	Absolute value (quantity/unit)	
	Output of unit product (quantity/unit)	
Method for achieving the above achievements		

Note Major pollutant refers to the principal pollutant having a significant impact on the local environment

Table 13.9 Public participation

How does the enterprise obtain the attention of society? What efforts are made to improve the enterprise's public image?	
How does the enterprise respond to society's key issue?	
If an enterprise has environmental performance exerting influences on the surrounding residents, would the residents be notified in advance? If any, what type of notice is provided?	
Has the enterprise been reported or petitioned due to an environmental problem? If any, please list the cause and frequency	
Is the enterprise in an environmental dispute currently? If any, please describe the dispute	
What methods does the enterprise use to publish its environmental performance to the public (website, newspaper, TV program)?	

(II) An Enterprise's technical report

An enterprise's technical report includes three sections: a text report, photographs of activities, and audio–video recordings. The features of the enterprise's environmental protection should be highlighted by the photographs of activities and audio–video recordings. The text report should include the following: ① basic information concerning the enterprise, including the name of the organization, its geographic position and picture, scale of production, starting time of production, state of operation, environmental honors, and contributions in recent years; ② current output of manufacturing techniques and pollutants; this includes a process flow diagram of the enterprise; ③ contrast and describe how the enterprise satisfies each of the basic assessment conditions in the table General Conditions and Assessment Methods of China's Environment Friendly Enterprises; ④ the social responsibility to be performed by the enterprise, including the current achievements of environmental protection, commitment, and public satisfaction index in the aspects of future environment and social contribution; ⑤ relevant evidentiary materials including the business license for the enterprise, evidentiary material of the enterprise's green purchases, and relevant awards material; and ⑥ self-review of the enterprise's environment protection work.

(III) Report of an enterprise's review

The report of an enterprise's review is primarily to perform self-review grading based on the *General Conditions and Assessment Methods of China's Environmentally Friendly Enterprises* and to specify the current weaknesses and goals of the enterprise.

13.5.3 Review and Others

(I) Review procedure

First, China's environmentally friendly assessment center performs a preliminary review of the enterprise's application materials and, then, organizes an expert group to review the enterprises that pass the first review. After a thorough review by an expert group, the qualified enterprises have experts organized by China's environmentally friendly assessment center perform spot assessments and acceptance of the enterprise and form spot assessment opinions. For the enterprises that pass the spot assessment and acceptance, they are submitted by China's environmentally friendly assessment center to ACEF to approve whether the title of "China environmentally friendly ×× enterprise" is awarded to the application enterprise. After being approved by the ACEF, the review comments and final results will officially be provided to the application enterprise by the China environmentally friendly assessment center. Finally, the ACEF awards the title of "China environmentally friendly enterprise" to the enterprise that has passed the review and issues the certificate. The ACEF will regularly release an announcement relevant to the assessments of China's environmentally friendly enterprises, so that enterprises can receive the review status.

(II) Other matters

The establishment of an environmentally friendly enterprise is a very good promotion for the future development of the enterprise, and the enterprise that obtains the title of "environmentally friendly enterprise" will receive the following rewards: ① License, denomination and press conference as a "China environmentally friendly enterprise"; ② the enterprise or product is recommended to the entire society by the ACEF; and ③ the enterprise and products can be propagandized and presented on China's environmental publicity information network, ACEF proceedings, and relevant media.

Additionally, to ensure that the enterprise's development behavior can constantly and stably meet the requirement of "environmentally friendly enterprise," China's environmentally friendly assessment center will regularly recheck the enterprise and products that have passed the review, to ensure that they consistently maintain the requirements of the environmentally friendly enterprise assessment standards. If an enterprise fails to pass a recheck by the ACEF three consecutive times, the title of "China environmentally friendly enterprise" would be canceled, and the canceling would be announced through major domestic media. For example, the important water pollution event in the Songhua River in 2015, through investigation, was found to be the result of an explosion that occurred in a workshop of the di-benzene plant of the Jilin Petrochemical Corporation. Approximately 100 t of benzene homologs and derivates (e.g., benzene, nitrobenzene) flowed into the Songhua River, resulting in severe pollution of the river's water. After the event, the title of "national environmentally friendly enterprise" was rapidly revoked from the Petro China Jilin Branch.

13.6 Case Analysis of Jinan Iron and Steel Co., Ltd.

13.6.1 Overview of Establishment Activities

(I) Basic status of the enterprise

Jinan Iron and Steel Co., Ltd. (hereinafter referred to as Jinan Iron and Steel) is the core enterprise of the Jinan iron and steel group corporation, and it is located in Industrial Park, Wangsheren Town, Eastern Suburb, Jinan City. Shandong (Jinan) Iron and Steel Group Co., Ltd. was established in 1958, its existing steel output is ranked sixth in our country, and it performs the entire manufacturing system from coking, raw material, sintering, pelletizing, ironmaking, and steelmaking to steel rolling. With middle plate, medium plate, hot rolled sheet, cold-rolled sheet, round steel, and deformed steel bar as its main products, it is the largest manufacturer and export enterprise for medium plate.

(II) Basis of environment protection work

As early as 2003, Jinan Iron and Steel passed the certifications of three systems, specifically, the ISO 9000 quality management system, ISO 14000 environment management system, and an occupational health and safety management system. In 2005, the National Development and Reform Commission, State Environmental Protection Administration, and six other ministries and commissions confirmed that Jinan Iron and Steel is one of the first of the pilot enterprises of a circular economy throughout the country. In 2008, the company was awarded the honorary titles of national environmentally friendly enterprise, national greening model unit, and environmentally friendly enterprise in China's steel industry.

(III) Objective of establishment activities

Facing the pressure of aggravating resources and environmental constraints, accelerating the transformation of Jinan Iron and Steel development methods, and promoting the strengthening of Jinan Iron and Steel are urgent requirements for achieving industry sustainable development. In accordance with the development of the internal and external standards, an "accurate, efficient, and harmonious" development strategy is prepared by Jinan Iron and Steel, and the purpose is to develop Jinan Iron and Steel into a domestic first-class and international advanced enterprise for sheet materials with high technical content and additional value.

(IV) Beginning of the establishment activities

In May 2004, Jinan Iron and Steel Group held a startup meeting to establish a national environmentally friendly enterprise. The group president released the relevant decisions and established the lead group and working team, read the rectification order for the establishment of a national environmentally friendly enterprise, and prepared the relevant remediation action, self-examination/self-assessment standards, which was the official launch by Jinan Iron and Steel to become a national environmentally friendly enterprise.

(V) Management of the establishment activities

During the establishment of the environmentally friendly enterprise, Jinan Iron and Steel employed the management mode of "four-level control," which consists of an expert, chief executive, safety, and environmental protection office and the second-level enterprise unit in the group, and the "three-level review," which consists of an enterprise's internal audit, a self-review by the chief executive, and an external review by experts. Documents are assembled from each activity of the enterprise and are issued to supervise and urge its rectification, to achieve multiple-aspect control from the company's superiors to its subordinates and inside to out.

13.6.2 Emphasis of the Establishment Activities

During the establishment process between 2004 and 2007, focusing on pollution regulation, implementation of cleaner production, and development of a circular economy, the following activities were implemented by Jinan Iron and Steel.

(I) Waste recycling

Between 2004 and 2007, Jinan Iron and Steel identified and completed nearly 100 continuous improvement projects. These include the comprehensive utilization project for production wastes, e.g., coke pushing and dust removal project in the

No. 3, 4, and 5 coking plants; finished product dust removal project in the No. 1 sintering plant and stock bin dedusting improvement of the pelletizing plant; wastewater comprehensive utilization project, e.g., living water treatment project and blast furnace gas washing backwater system project of the No. 1 iron mill; slag-washing water improvement of the No. 1–6 blast furnaces; backwater system improvement project of the blast furnace area; tail-end project of industrial water zero release; and waste heat and energy recovery project, e.g., turbid circulating water system cooling transformation project of the No. 1 blast furnace and the waste heat steam condensate recovery project. In these projects, mineral powder, coke and other fly ash are used as sinter burdening, high-grade steelmaking dedusting sludge pelletizing is directly used for ironmaking production, and the low-grade steelmaking dedusting sludge is sent to a mining company for selection or producing iron ore, which achieves the resource transformation of smoke/dust and other wastes. Through waste recycling, Jinan Iron and Steel cannot only reduce their resource utilization and waste discharge but can gain considerable economic benefits. In 2006, 798,000 t of dedusting ash was collected by Jinan Iron and Steel, including 419,000 t of iron-bearing dust, 287,000 t of coke powder, 88,000 t of steelmaking and ironmaking sludge and nearly 5000 t of lime powder, achieving approximately 120 million Yuan of annual profit.

(II) Implementation of cleaner production

Implementation of cleaner production is the core of establishing an environmentally friendly enterprise, and cleaner production is implemented by the Jinan Iron and Steel in the following manners:

Cleaner processes are selected. To save energy and reduce consumption, extract potential, and increase benefits and by employing imitative innovation, "four-full and one-spray" is implemented by Jinan Iron and Steel, specifically, full ironmaking clinker, full ironmaking refining, sequence casting, one-heating forming of steel rolling and improvement of the coal injection rate of the blast furnace, to complete the energy conservation optimization of the enterprise's manufacturing technique. In this manner, Jinan Iron and Steel eliminates the lagging process, shortens the technological process, improves the intensification degree of the production organization, and achieves the preliminary optimization of an energy-saving full flow structure.

Clean resources are selected. Adhering to the policy of beneficiated burden material, "three-high and two-low" is implemented by Jinan Iron and Steel, namely high blast temperature of ironmaking, high coal powder injection, high coefficient, low delay ratio, and low coke ratio, so that the enterprise can exclude the unavailable non-metallic impurity, poisonous and harmful elements from the smelting procedure as much as possible and can prevent the wastes and inferior-quality products before the steel rolling process. The key point of the policy of beneficiated burden material lies in the prior iron smelting; work should begin from the source to optimize mineral occurrence, structure the ore blending, coking, and sintering processes, improve the furnace entry quality, and reduce the amount of slag. The consumption of energy and resources can be reduced from the source to reduce waste production. The policy of beneficiated burden material should extend to the post iron smelting process, primarily to reduce the consumption of steel-making lime and iron and steel materials, ensuring billet hot charging via zero-defect casting blanks and providing high-quality blanks for steel rolling.

Cleaner production techniques are employed. More than 95% of the water for an iron and steel enterprise is used for process cooling, environment protection, and dedusting. The innovative thinking of "classification by water quality, small-radius circulation, and regional closure" is employed to industrial water by Jinan Iron and Steel, and according to the different technological characteristics of each procedure, different water quality and water temperature processing techniques are used to digest the sewage in process, obtaining obvious effects. The sewage treatment plant in the living quarters was built by Jinan Iron and Steel while strengthening the comprehensive treatment of industrial water. Its daily processing capacity has reached 7200 t, and after being processed, it is used for production and greening.

(III) Development of a circular economy

The Jinan Iron and Steel circular economy has three stages: the first is employing latent potential and energy-saving and cost-reduction based on cost; the second stage is the cleaner production phase guided by a focus on product and quality; and the third stage is the initial stage of the circular economy guided by a focus on resources and the environment. The three stages increase from low to high, allowing enterprise improvement under the guidance of a circular economy concept to be implemented gradually. It primarily includes the aspects below.

Technological innovation—Innovation input is constantly strengthened by Jinan Iron and Steel. Between 1996 and 2005, more than 120 energy-saving clean processes were independently researched and developed to form 117 patents, receiving more than 77 scientific and technical progress rewards from the ministerial and provincial levels. It primarily includes a dry quenching technique, a wet-milling coke powder coking blending technique, steam technology replaced by a coking process thermal-conduction oil, and a slag water quenching technique. The study and application of the dry quenching technique and the efficient and gradient utilization of thermal energy resources achieved their second scientific and technical progress reward. Technological innovation is a key strength for Jinan Iron and Steel to develop a circular economy.

Management innovation—With a core of efficient utilization of resource energy, market and price mechanisms are used by Jinan Iron and Steel to constantly improve and perfect their relevant rules and regulations, to effectively guide each organization to cherish resource energy and promote efficient utilization. For example, in the management of water resources, the prices of new water, clean circulating water, and turbid circulating water are 4.5, 0.5, and 0.35 Yuan/m^3, respectively. Additionally, the innovation incentive mechanism is constantly improved to award a large prize to the staff member making an outstanding contribution in innovation work, which provides positive support to solving the enterprise's key technical problems and to break through the "bottleneck" restriction. By referencing the international advanced management experience, an excellent performance management model is introduced to improve the enterprise's comprehensive management level and to promote the efficient utilization of resource energy and a constant reduction of the environmental load.

Efficient utilization of resources—The policy of beneficiated burden material reduction is implemented by Jinan Iron and Steel to maximize the reduction of mixing non-metallic impurity into the smelting procedure, to explore refining and finish rolling, and to improve the steel rolling yield. Between 2000 and 2005, the steel rolling yield improved by 4.43%. Focusing on industrial water, the efficient utilization mode of "supply based on quality, grading treatment, small-radius circulation, and regional closure" is implemented to optimize water use, ease the pressure of tail-end treatment, and reduce water consumption.

Efficient transformation of energy sources—Through process improvement and structure optimization, efficient transformation of energy sources is achieved by Jinan Iron and Steel. For example, in the pipeline steel project of a national key project, the new process of the billet online direct delivery from a 120 t converter to medium plate production line and the process of directly sending a high-temperature billet into a pipeline steel heating furnace are innovated to greatly improve the thermal energy conversion efficiency of the process. Another example is the first domestic dry quenching device built by Jinan Iron and Steel through independent innovation to change the previous water quenching method, reducing the waste of resources and energy and environmental pollution and achieving the "four-improvement" of heat conversion efficiency, coke quality, environmental quality, and economic benefits. The structure optimization of the full-process energy system is referenced to implement the coal gas-saving methods of "two-adjustment" and "three-full." The "three-full" method refers to continuous casting efficient full-hot charging/loading, efficient full-heat accumulation of a heating furnace, and full-scale efficient rolling of medium plate; "two-adjustment" refers to "gas structure adjustment" and "gas calorific value adjustment." Through 2005, dry quenching power generation, gas–steam combined circulation power generation, and blast furnace top-pressure recovery turbine (TRT) power generation have reached a total of 730 million kWh.

Efficient regeneration of metabolites—With the slag, dust, water, steam, and other intermediate emissions produced in the iron and steel production process as resources, the enterprise's metabolic waste is decentralized by Jinan Iron and Steel by implementing "four-closed circuit," including gas, slag, and metallurgical ferric dust, industrial water and waste heat steam closed circuits, to achieve spot governance, spot transformation and spot use, which not only requires a low investment, has a quick effect and is simple to implement but also has a low operating cost, reduces pollution and increases benefits. For example, focusing on the governance concerns of metallurgical slag and ferric dust, the slag quenching process is researched and implemented by Jinan Iron and Steel to mix water-quenched slag into sintered material, to achieve the full utilization of the metallurgical slag and ferric dust.

13.6.3 Effect of the Establishment

Since implementation as an environmentally friendly enterprise in 2004, Jinan Iron and Steel has increased their level of investment to expand capacities without increasing emissions, improved their environmental index annually, and constantly improved air quality, achieving their best benefits and finest environment in 2006.

(I) Environmental and economic benefits

In 2006, by establishing an environment optimization enterprise, the comprehensive energy consumption per ton of steel in Jinan Iron and Steel reached 603 kg of standard coal, and compared with the previous year, the 10,000 Yuan of added value energy consumption reduced by 3.86% and the total amount of energy conservation was 265,000 t of standard coal. This amount is lower than the average value (645.12 kg standard coal/t) of the comprehensive energy consumption per ton of steel in the major large- and middle-sized iron and steel corporations. In the first half of 2007, the comprehensive energy consumption per ton of steel of Jinan Iron and Steel reduced to 593 kg of standard coal, making it the world leader.

Fifteen of the 17 major production processes of the entire company reached special-grade process energy consumption standards in 2006; the second-class and lower processes were eliminated. The technique of steelmaking with less energy consumption is achieved by the number one and number three steel companies. In the entire company, the 10,000 Yuan added value energy consumption reached 6.316 t of standard coal, a reduction of 0.235 t from the previous year, a reduction rate of 3.6%. The total amount of energy consumed throughout a year can reach 250,000 t of standard coal, which exceeds the responsibility goal. Specifically, 133,300 t of standard coal annual energy consumption was signed with the provincial government. The comprehensive new water consumption per ton of steel reduced to 3.61 m^3, decreasing 0.26 m^3 from the previous year. Residual heat and energy is utilized to generate 1.09 billion kWh of electricity, an increase of 440 million kWh, a 69% improvement over the previous year.

During the establishment of the environment optimization enterprise, the emission amounts declined annually. Specifically, the discharge of waste water decreased to 14.98 million t in 2006 from 18.95 million t in 2000, a decrease of 21%; and water discharge per ton of steel decreased to 1.34 t in 2006 from 5.7 t in 2000, a decrease of 76%. The discharge of suspended solids, COD, and petroleum in waste water decreased to 889 t, 940 t, and 43 t, respectively, in 2006 from 2637, 1218, and 127 t in 2000, decreases of 66, 23, and 66%, respectively. Smoke (dust) discharge of exhaust gas decreased to 7879 t in 2006 from 17,969 t in 2000, a decrease of 56%, the dust discharge per ton of steel decreased to 0.63 kg in 2006 from 5.4 kg in 2000, a decrease of 88%, and the SO_2 discharge per ton of steel decreased to 1.05 kg in 2006 from 2.0 kg in 2000, a decrease of 47%. Reduction of the environmental load greatly reduced the pressure on the environment and improved environmental quality. Air quality improved annually. In 2006, in comparison with 2000, the annual average of the air quality indexes within the plant area, specifically, SO_2, NO_2, total particulate matter, CO, and atmospheric dust, decreased 21, 37, 49, 72, and 46%, respectively.

Substantial progress has been achieved in the expansion of circular economy technology. The dry quenching technique has been exported to seven domestic and international enterprises, such as India and Guangxi Liuzhou Iron and Steel (Group); the sintering power generation technique has been exported to four companies, such as the Jiangsu Shagang Group and Huaisteel Special Steel Co., Ltd., and the saturated vapor power technology has been exported to Tangshan Yanshan Iron and Steel. Many techniques, including coal moisture control, slag hot disintegrating, and waste heat recovery of heating furnace, have encouraging expansion prospects. The development of a circular economy is important to reduce cost and ease environmental stress.

(II) Establishing characteristics of an environment optimization enterprise

One of the characteristics is to form a circular economy guided by the steel-power co-production of Jinan Iron and Steel. The "three-integrated and two-adjustment" policy is implemented in the enterprise (specifically, the continuous casting efficient full-hot charging/loading, efficient full-heat accumulation of heating furnace, full-scale efficient rolling of medium plate, and gas structure adjustment and gas calorific value adjustment).

By saving residual gas, the LHV mixed gas was regarded as fuel to build a gas–steam combined cycle power generation phase I project in 2004. In 2006, the phase II project began production. The installed capacity of the two phases is 543 MW, and the annual electric energy production can reach 3.13 billion kWh. Gas–steam combined cycle power generation is free from exhaust gas, waste residue, and pollution, and is capable of reducing 3 billion m^3 of blast furnace gas emission and 0.7 billion m^3 of coke oven gas yearly, and eliminating 145,000 t of SO_2 and 1.5 million t of CO_2 emission annually. During the beginning of 2007, the gas–steam combined cycle power generation of Jinan Iron and Steel was officially registered successfully in the United Nations; the project was the first successfully registered gas–steam combined cycle power generation enterprise in China's metallurgical industry. The CO_2 certified emission reduction per year is approximately 1.5 million t, making it the project with the maximum CO_2 emission reduction in gas–steam combined cycle power generation within the metallurgical industry. In addition, Jinan Iron and Steel has successfully built steelmaking steam and sintering machine waste heat power generation, developing techniques as part of a domestic initiative.

Through the establishment activities of an environment optimization enterprise, Jinan Iron and Steel recognizes the role of iron and steel enterprises in society and the natural environment from the perspective of social responsibility and value innovation. Through constant technological development and function evolution, Jinan Iron and Steel verifies that the iron and steel enterprise can develop into an environmentally friendly enterprise with cleaner production, green manufacturing, and the ability to digest its own and social wastes from emission pollutions.

Furthermore, as an important link in the social circular economy chain, and with its own unique superiority, an iron and steel enterprise can further achieve community-friendly and socially friendly conditions, to achieve harmony and unity among enterprise revenue, social benefit, and environmental benefit.

Classroom Discussions and Assignments

Topic Discussion

In this course, the basic method and review procedure for assessment are discussed and combined with the historical practice for assessment of an environmentally friendly enterprise in our country. As previously mentioned, with the gradual implementation of the assessment activities, several hundred environmentally friendly enterprises have been built, and an environmentally friendly enterprise alliance has been formed. Please discuss the following topics:

(1) Investigate and survey the development status of the environmentally friendly enterprise alliance in our country, and propose improvement ideas and suggestions;
(2) Select a specific familiar enterprise for social practice activity, understand the difference between the enterprises' situation and an environmentally friendly enterprise, and propose improvement ideas and suggestions;
(3) Specific to a certain existing environmentally friendly enterprise, implement a social practice activity to investigate, and survey the key emphasis and results of establishing an environmentally friendly enterprise;
(4) What are currently the disadvantages of establishing and managing an environmentally friendly enterprise in our country? How can they be improved?

Homework and Presentation in Class

For this topic, students are required to leave the classroom and go to a site to administer a survey; this will allow students to comprehensively apply the knowledge that they have learned in the classroom to identify a problem that exists in an actual enterprise operation, evaluate the problem, and attempt to solve it. This assignment is a challenge and requires innovation. Henceforth, the homework will be used to train students' ability to apply integrated knowledge and innovative thinking; students should think positively and propose their own ideas and suggestions. The task of this lesson is to choose one of the above topics and write an investigative report on the basis of the survey for sharing, using PowerPoint, in the class via the group and to summarize it into a 3000-word essay. The grading scale is shown in Table 13.10.

Table 13.10 Grading of the group report in class

Report title	Logic (maximum 10 points)	Ability to apply integrated knowledge (maximum 10 points)	Innovation (maximum 10 points)	Presentation skill (maximum 10 points)	Total points

Recommended and Referenced Literatures

1. Establishment of the Resource-saving and Environment-friendly Iron and Steel Enterprise Editorial Board. 2006. Establishment of the resource-saving and environment-friendly iron and steel enterprise. Beijing: Metallurgical Industry Press.
2. Chazhong, Ge, and Jiang Hongqiang. 2010. *Industrial pollution control and environment friendly enterprise*. Beijing: China Environmental Science Press.
3. State Environmental Protection Administration. 2003. Notice for holding the state environment friendly enterprise creating activity [No. 92 H.F. (2003)]. [EB/OL]. (2003-05-23) [2016-08-09]. http://www.zhb.gov.cn/gkml/zj/wj/200910/t20091022_172194.htm.
4. Pollution Control Division of Environmental Protection Department, Environmental Development Centre of Environmental Protection Department. 2009. Work manual of the establishment of national environment friendly enterprise. Beijing: China Environmental Science Press.
5. Jingjiang, Li. 2006. *Enterprise' green operation-the only way of sustainable development*. Beijing: Tsinghua University Press.
6. Fanje, Liang. 2005. *Corporate environmental management: corporate environmental management and administrative approval*. Beijing: China Environmental Science Press.
7. Zuomiao, Xie. 2008. *Environmentally friendly economic development mode*. Beijing: Minzu University of China Press.
8. Yinghua, Zhang, and Zhang Jianyu. 2010. *Theory and practice of enterprise's environmentally friendly operation*. Beijing: Economic Science Press.

14 Enterprise Environmental Management

14.1 Presentation of Core Issues

From the previous discussion, we know that a sustainable development enterprise should be able to constantly provide high-quality services to society and receive economic benefits during the service process and, simultaneously, continuously improve environmental quality during its business activities. Under the existing environment "bottleneck," the improvement of environmental performance should be especially emphasized in the enterprise's development; therefore, an environmentally friendly enterprise becomes the key emphasis during this stage of an enterprise's improvement. The establishment of an environmentally friendly enterprise not only must consider the product (or service) produced by the enterprise and reduce the environmental impact of its entire life cycle as much as possible (as determined by the LCA assessment method provided in Chap. 11), but it must avoid poor service caused by the product prior to it being placed into production (as determined by the eco-design measures provided by DfE in Chap. 12) during the product conception stage. Additionally, the effect of the enterprise's environmentally friendly construction must be determined by means of an assessment (the assessment method for an environmentally friendly enterprise was provided in Chap. 13); however, knowing how to improve its environmental performance does not indicates an action that an enterprise put the measures to practice for improving environmental performance voluntarily and initiatively. If necessary, an enterprise must be brought onto the right direction of performing environmental improvement by means of business management measures. Therefore, the topic of this chapter is methods to perform an enterprise's integrated management.

The management of an enterprise will be involved in the relevant issues, such as who is responsible for management, what is the content of management, and how should management be performed? In this chapter, we will discuss the required qualifications of the environment administrator, management changes in the content and scope relative to traditional business management, the universal international

J. Mao et al., *Circular Economy and Sustainable Development Enterprises*,
https://doi.org/10.1007/978-981-10-8524-6_14

method of environmental management, and the existing major management measures.

14.2 What Are the Management Requirements of an Enterprise?

As the resource transformation organization, an enterprise is the interface between human beings and the environment, and its environmental performance will directly influence the environmental quality. The environmental management of the enterprise is not only the management of pollutants but is also the management of resource protection; it is an effective method to radically solve the resource and environmental problems. An environmentally friendly improvement to the enterprise is an important phase in its environmental management.

14.2.1 General Requirements of Business Management

To achieve an environmentally friendly enterprise, the practice of unilaterally pursuing economic benefit must be changed, and an objective of environmental protection must be established. With the entire life cycle of a product or service as the management scope, and based on participation of the entire staff, the synchronous and continuous development of the economy and the environment via preventing pollution and perfecting an environment management system can be achieved, to achieve mutual benefits among producer, consumer, and society. This type of management method is called total environmental management (TEM).

An enterprise is requested to make changes in many areas, such as management objectives, perceptions, management functions, and management methods by TEM, including the following activities.

(1) Change in perception—To change economic benefit emphasis into equal emphasis on economic benefit, society, and the environment, the impact of the enterprise's operation on the external environment and its contribution to society must be considered along with the economic benefit sought from the enterprise's operation. To change from pollution regulation into pollution prevention, the enterprise transfers the "afterward treatment" of the environmental problem into "beforehand prevention" by placing the emphasis on pollution prevention. In this manner, "pollution control" is transformed into "pollution prevention." This establishes the concept of social dedication, which integrates an enterprise TEM system into the historical mission of benefiting human beings and their descendants.
(2) Full participation in environmental management—The content of environmental management is dispersed into each functional department, and post, all staff members explore the measures for pollution prevention and reduction

from their own work perspective and bring their measures into the integrated environmental improvement strategy of the enterprise. Meanwhile, the environmental responsibility system of each functional level of personnel is prepared, with a clear task, function, duty, and right and stipulates the environmental objective, indicator, and plan, so that each department performs its own functions and acts in close coordination, forming a close work system for environmental management.

(3) Comprehensive application of multiple management methods—To establish and perfect the data information system, quantify and objectively check the business operation levels to employ the ISO 14000 environment management system for management according to the plan–execute–inspect–process cycle, and constantly apply new science and technology achievements into the enterprise's environmental management, e.g., instrumentation and online monitoring and management.

14.2.2 Quality of the Enterprise Manager

Enterprise manager refers to the person who coordinates an enterprise's manpower, material resources, and financial resources to achieve the established goals via planning, organization, control, motivation, and leadership. An enterprise manager has the right to allocate the resources demanded for the enterprise's development and decides the development direction of the enterprise. He or she confirms the enterprise's development goal, frames the plan, and is the work instructor, an important role in the enterprise.

Degradation of the external environment causes great challenges to the enterprise's development and also provides an opportunity for the enterprise manager to reform the business management mode. It evolves into a situation of satisfying the environmental change requirement and seeking durable competitive advantage from the traditional pattern of managing internal resources and seeking the optimal economic output. In this case, an enterprise manager should not only possess the ability to control manpower, finance, materials, and other resources, but should look forward, establish lofty goals and have the ambition to guide the enterprise to protect the environment, to improve the enterprise's environmental performance and make it an environmentally friendly enterprise. At this time, the enterprise manager should directly support and encourage the implementation of environmental management measures for management, organization, and capital investment or should specifically appoint an environmental administrator to oversee environmental matters involving various activities, such as enterprise operation and production. Undoubtedly, the qualities of the enterprise manager and the environmental administrator are crucial to the enterprise's environmental performance.

To build an environmentally friendly enterprise, the enterprise manager and environmental administrator should possess the following basic qualities:

(1) Sense of mission to encourage the sustainable development strategy—The enterprise manager should fully approve the sustainable development strategy and should regard the environmentally friendly construction of the enterprise as a necessary mission.
(2) Management and leadership ability—This refers to the influence of the manager, including authority and non-authority power. He or she not only must restrain the personnel behavior norm via the enterprise's rules and regulations but must also promote staff self-awareness of environmental protection via quality training, to enhance the executive force of the enterprise manager.
(3) Learning and innovation ability—The enterprise manager must constantly improve his or her ability to cope with internal affairs, including adapting to the requirements of social progress, economic development, market change, other human factors, and the requirement of resources and environment change for the enterprise. An enterprise manager is requested to constantly study to encourage the enterprise's technical progress via innovation.
(4) Teamwork—The manager should be capable of impelling, driving, and influencing the entire team to achieve the stated objective instead of achieving only the enterprise objective.
(5) Self-control—In the face of temptation and pressure, the manager must be able to correctly treat and address the relationship between economic benefit and environmental benefit.
(6) Acute strain capacity—The manager must understand the changes in external politics, economy, law, and society to avoid weaknesses and take full advantage of good opportunities to promote enterprise development.
(7) Perseverance and willpower—Usually, there are risks, opportunities, and many difficulties during an enterprise's development, e.g., commercial, market, policy, and credit risk, which require the manager to maintain clear cognition and actively employ measures to advance the stated objective bravely and firmly.

14.2.3 Emphasis on Environmental Management

For the environmental management of the enterprise, determine what type of environmental problems would be confronted. How do these environmental problems form? Which industry or field should be the focus of attention? What is role the to-be-managed enterprise playing? By means of the environmental pollution problem, we provide examples to solidify the process of environmental management emphasis.

(I) Example 1: determine the key pollutant and field via a general investigation of pollutants

Pollution prevention and control is one of the key factors in environmental management. What are the pollutants? Which department is their source? A general investigation of pollutants is an important method to clarify these problems.

In 2007, the first national investigation of pollutant sources was implemented in our country to clarify the emission behavior of various pollutants in China. According to the results of the general investigation, the key pollutants and their discharge in our country were the following: in wastewater, COD, 30.2896 million t; ammonia nitrogen, 1.7291 million t; heavy metals, 900 t; TP, 423,200 t; and TN, 4.7289 million t; and in exhaust gas, SO_2, 23.2 million t; NOx, 17.977 million t; smoke, 11.6664 million t; industrial dust, 7.6468 million t; industrial solid wastes, 49.1487 t; and industrial hazardous waste, 394,000 t. Additionally, the source industries of the key pollutants in our country were also determined: COD and ammonia nitrogen were discharged from eight industries, such as papermaking and spinning, comprising 83 and 73% of the total amount of industrial discharge, respectively. The SO_2 and NOx discharges of six industries, e.g., electric heating and nonmetallic mineral products, comprise 89 and 93% of the total amount of industrial discharge, respectively. On this basis, a series of pollution prevention and control measures are purposefully implemented in our country to improve the overall environment. This also proves that a pollutant general investigation can overcome the disadvantages of the former environmental statistical survey and bring opportunities for the establishment of a scientific and efficient environmental statistical system. It is reported that the Ministry of Environmental Protection will perform the second national survey of pollutants between 2017 and 2018.

(II) Example 2: screening of the key industrial sectors

In regional environmental management, the relationship between economic development and environmental protection must be balanced, and these two factors are inseparable from the local residents. In this case, the concerning social economy and environmental protection matters may be selected by means of expert review, public participation, and other mechanisms. Each matter is empowered and graded, and then, the relevant industry sectors are ranked to determine the key industrial sectors.

For example, in the ecological planning to screen the key industrial section of a specific basin through survey, three major influential factors, namely economic output, water consumption, and wastewater and its pollutant discharge, are selected, and the weights of the above three elements are determined as follows: economic output, 0.3; water consumption, 0.3; and wastewater and its pollutant discharge, 0.4 (wastewater discharge, COD, and ammonia nitrogen occupy 0.13 of each). Each industrial sector is scored and weighted to determine the top ten industrial sectors within this basin, as shown in Table 14.1. The top ten industrial sectors occupy

Table 14.1 Example of key industrial sector screening

	Total output value/hundred million Yuan	Water withdrawal/10^6 t	Discharge amount of wastewater/10^6 t	COD discharge/Ten thousand t	Ammonia nitrogen discharge/t
A specific sub-basin	448.60	106.41	67.35	1.77	96.05
Papermaking and paper products industry	39.34	36.09	29.17	1.62	0.08
Chemical raw material and chemicals manufacturing	30.02	5.83	2.05	–	8.53
Transport and communication facilities manufacturing industry	80.27	2.86	2.44	–	–
Textile industry	30.36	22.97	17.85	–	–
Electric apparatus and equipment manufacturing industry	74.32	1.98	1.03	–	–
Leather, fur, feather (down feathers), and their products industry	13.58	9.75	7.67	–	57.02
Manufacture of rubber	13.38	1.26	0.95	–	–
Nonmetallic mineral products industry	18.72	0.42	0.15	–	–
Manufacture of chemical fibers	7.05	0.69	0.28	–	–
Communication device, computer, and other electronic equipment manufacturing	25.66	0.57	0.48	–	3.90
Total quantity of the top ten industries	332.7	82.42	62.07	1.62	69.53
Proportion of the top ten industries (%)	74	77	92	92	72

75% of the industrial output, water consumption, and ammonia nitrogen discharge, while in wastewater and COD discharge, they occupy 92%; it is an important industrial sector for environmental management in this basin.

14.3 Method to Perform Standardized Management Within the Enterprise

14.3.1 Overview of an Environment Management System

A sound environment management system is an important guarantee to effectively solve many problems. To promote the overall improvement of global environmental quality, ISO has successively issued the international environment management system ISO 14000 series of standards. Their purpose is to strengthen the organization of the company's or enterprise's environmental awareness, management ability, and safeguarding measures.

In ISO 14001, the environment management system (EMS) is defined as the constituent portion of the overall management system in an organization (i.e., a company or an enterprise). It includes the institutional framework, planning activity, organizational responsibility, convention, procedure, process and resource demanded by the preparation, implementation, realization, review, and maintenance of the environmental policy; it also includes the contents of an organization's environmental policy, objectives, and indicators. The EMS is a management tool in an organization. The purpose of implementing ISO 14000 in an enterprise and establishing EMS based thereon is to reduce the environmental pollution caused by various activities, conserve resources, and promote the society's and enterprise's sustainable development. Its significance includes the following:

(1) Providing the "green passport" for international trade indicating the implementation of the ISO 14000 standard;
(2) Improving an enterprise's image, reducing environmental risk, gaining advantages in market competition, and creating business opportunity;
(3) Improving management ability, forming systemic management mechanisms, and perfecting the overall management level of the enterprise;
(4) Grasping the environmental conditions, reducing pollution, and reflecting the concept of "cleaner production";
(5) Saving energy and reducing consumption, lowering cost, reducing various environmental costs, and gaining significant economic benefit; and
(6) Conforming to the national policy of "sustainable development," free from the constraint of domestic and overseas environmental requirements, enjoying domestic and overseas environmental preferential policy and treatment, promoting harmonious and sustainable development between an enterprise's economic development and environmental improvement.

The implementation of the international EMS in our country is valued by the Chinese government. Since 1996, we have implemented the certification activities of the EMS pilot, and currently, ISO 14000 EMS standard certification has been implemented by more than 10,000 enterprises, effectively reducing resources and energy consumption, decreasing the enterprise's production cost, and protecting the environment.

14.3.2 ISO 14000 Standard Series

The existing EMS international standard is shown in Table 14.2. It includes environmental management system (EMS), environmental auditing (EA), environmental label (EL), and many other topics within the field of environmental management and systematically stipulates all the terms used in environmental management, EA methods, and presentation of assessment results. ISO 14001 and ISO 14004 are the basic standards, while ISO 14010 through ISO 14049 are supporting technical standards. The ISO 14000 series of standards is an important basis to instruct various enterprises and companies to obtain correct environmental performance.

In Table 14.2, ISO 14001 and ISO 14004 are the basic standards, the final ISO 14063, ISO 14001, and ISO 14004 are the primary standards, and the others are supporting technical standards.

14.3.3 Examples of International Standards Content

ISO 14001, *EMS Specification and User Manual*, is the core of the ISO 14000 series of standards, and it requires an organization to support environmental protection, prevent pollution and constantly improve by building an environment management system and, furthermore, to prove EMS's conformity and environmental management level by obtaining the approval of a third party's certification authority. The EMS requirement described in chapter IV of ISO 14001 is the core content of the system, and its basic structure is as shown in Table 14.3. In the table, we choose several important clauses to describe and provide examples of their implementation.

(I) Example 1: environmental factor

1. Standard clause

An environmental aspect is the element's interaction with the environment within the activity, product, and service (including those planned or newly developed) of an organization. The clause 4.3.1 regulations on the environmental aspect of ISO 14001 are as follows: The organization should establish, implement, and maintain one or more procedures to identify the environmental aspect capable of being

Table 14.2 List of existing business management standards

Standard category	Standard number	International standard no.	Standard name	National standard no.
Environment management system (EMS)	14001–14009	ISO 14001:2004	EMS—requirements and user manual	GB/T 24001—2004
		ISO 14004:2004	EMS—general guidance for principle, system and support technology	GB/T 24004—2004
Environmental auditing (EA)	14010–14019	ISO 14010:1996 ISO 14011:1996 ISO 14012:1996	Guideline of the quality and (or) EMS review	GB/T 19011—2003/ ISO 19011:2002
		ISO 14015:2001	Environmental management—environmental assessment of sites and organizations (EASO)	GB/T 24015—2003
Environment label (EL)	14020–14029	ISO 14020:1998 (first edition) ISO 14020:2000 (second edition)	Environmental management—EL and announcement—general principles	GB/T 24020—2000 (First edition)
		ISO 14021:1999	Environmental management—EL and announcement—self-environmental announcement (Type-II EL)	GB/T 24021—2001
		ISO 14024:1999	Environmental management—EL and announcement—Type-I EL—Principles and procedures	GB/T 24024—2001
		ISO 14025:2006	Environmental management—EL and announcement—Type-III EL—Principles and procedures	GB/T 24025—2009
Environmental performance evaluation (EPE)	14030–14039	ISO 14031:1999	Environmental management—EPE manual	GB/T 24031—2001
Life cycle assessment (LCA)	14040–14049	ISO 14040:2006	Environmental management—LCA principles and framework	GB/T 24040—2008
		ISO 14044:2006	Environmental management—LCA requirements and guidelines	GB/T 24044—2008
		ISO/TR 14047:2003	Environmental management—LCA ISO 14042 application	GB/T 24025—2009
		ISO/TR 14049:2000	Environmental management—LCA ISO 14041 application	GB/T 24025—2009

(continued)

Table 14.2 (continued)

Standard category	Standard number	International standard no.	Standard name	National standard no.
Terms and definitions (T&D)	14050–14059	ISO 14050:2009	Environmental management—terms	GB/T 24050—2004
Others	14060–14100	ISO/TR 14062:2002	Environmental management—design and development for bringing environmental factors into products	GB/T 24062—2009
		ISO 14063:2006	Environmental management—environmental information—guidelines and examples	GB/T 24025—2009

Table 14.3 Structure of the contents of EMS ISO 14001

First-level clause	Second-level clause	Examples
4.1 General requirements		
4.2 Environmental policy		
4.3 Planning	4.3.1 Environmental factor	Example 1
	4.3.2 Environmental regulations and other requirements	
	4.3.3 Objective, indicator and program	Example 2
4.4 Implementation and operation	4.4.1 Resource, function, responsibility and authority	Example 3
	4.4.2 Ability, training and awareness	
	4.4.3 Exchange of information	
	4.4.4 Document	
	4.4.5 Document control	
	4.4.6 Operational control	
	4.4.7 Contingency reserve and response	Example 4
4.5 Inspection	4.5.1 Monitoring and measurement	
	4.5.2 Conformity assessment	
	4.5.3 Inconformity, corrective and preventive measures	
	4.5.4 Record of environmental governance process	Example 5
	4.5.5 Internal check	
4.6 Management review		

controlled or exerting influence on the activity, product, and service within the EMS coverage area; additionally, the organization should consider the included plan or new development, new or repaired activity, product and service, and other factors. The organization should determine the factor that exerts or might exert significant impact on the environment (namely an important environmental aspect). The information should be captured in a document and updated by the organization in a timely manner. The organization should ensure that the important environmental aspect is considered while building, implementing, and maintaining the EMS.

2. Example of the relationship between environmental aspect and environmental impact

In numerous environmental aspects, the important environmental aspect is emphasized, namely the one possessing or capable of exerting important environmental impact. The environmental aspect is closely related to the environmental impact; Table 14.4 shows the relationship between environmental aspect and environmental impact, to carefully distinguish the differences of the two in application.

3. Application example: Evaluation principle and list of the important environmental aspects in a specific company

Table 14.4 Relationship between environmental aspect and environmental impact

Environmental aspect	Environmental impact
Discharge of boiler's smoke	Air pollution
Wastewater discharge	Pollution of water resources
Coal consumption	Consumption of natural resources
Discharge of factory noise	Noise pollution
Waste oil-bearing duster cloth	Land pollution

Assessment of important environmental aspects is to determine the environmental aspects with significant or potentially significant impact to the environment from the identified environmental aspects via survey, thus identifying the important environmental aspect. First, the aspect with a wide scope of influence on the enterprise's image does not conform to environmental protection laws and regulations and other requirements and is of significant concern to the community that can be directly determined as an important environmental aspect. Second, by considering the scope of influence, incidence, frequency, regulatory compliance, social attention, economic ability, and other factors, we can determine the important environmental aspect via a scoring method. Table 14.5 is the list of the important environmental aspects of a specific machinery manufacturing company.

(II) Example 2: objective, indicator and program

The objective, indicator, and program requirements of environmental management are stipulated in clause 4.3.3 of ISO 14001.

(1) For each relevant internal function and level, the organization should establish, implement, and maintain the environmental objective and indicator of the documents.
(2) The objective and indicator should be measured. The objective and indicator should be in accordance with the environmental policy, including the commitments to pollution prevention, continuous improvement, and observance of applicable laws and regulations.
(3) The organization should consider the laws and regulations from their own environmental aspect. In addition, available technical solutions, finance, operation and business requirements, and other opinions of related parties should be considered.
(4) The organization should prepare, implement, and maintain one or more solutions used to achieve its objective and indicator, and the solution should include a stipulation of the responsibility to achieve the objective and indicator of each relevant function and level in the organization and the method and schedule for achieving them.

The environmental objectives, indicators, and plans of action for a specific machine manufacturing company in 2006 are shown in Table 14.6.

Table 14.5 List of the important environmental aspects of a specific machinery manufacturing company

No.	Important environmental aspect	Environmental factor/material composition	Related department	Activity/product/service	Environmental impact	Time	Status	Applicable laws, regulations, and clause numbers	Order of evaluation	Recommended control measures
1	Machine tool oil leakage	Engine oil and lubricating oil	Each machining workshop	Equipment leakage	Land pollution	Now	Normal	*Directory of National Hazardous Wastes* in chapter IX of the *Law on the Prevention and Control of Environmental Pollution by Solid Wastes of the People's Republic of China*	B	Environmental management program and operational control procedure
2	Solid wastes	Industrial refuse, production waste, and household refuse	Casting branch, boiler room, each workshop, and office area	Slag, cinder, smear metal, and organic wastes.		Now	Normal	Chapter IX of the *Law on the Prevention and Control of Environmental Pollution by Solid Wastes of the People's Republic of China*	B	Environmental management program and operational control procedure
3	Occurrence of fire and explosion accidents		Each workshop, oil depot, chemical engineering, paint store, acetylene oxygen bottle depositary			Future	Emergency	Articles 3–5, 18–21, 23–26, 30–33, 36, 38–40 and 45 of the *Administrative Provisions of Fire Protection of Organ, Enterprise and Public Institution*	A	Emergency preparedness and response control procedure and operational control procedure

(continued)

Table 14.5 (continued)

No.	Important environmental aspect	Environmental factor/material composition	Related department	Activity/product/service	Environmental impact	Time	Status	Applicable laws, regulations, and clause numbers	Order of evaluation	Recommended control measures
4	Reduction of production, domestic water and power consumption	Water and electricity	Water/power department for power reserves	Production water, electricity, domestic water, and living power utility	Water/power resource consumption	Now	Normal	Articles 1, 2, 3, 5, 6, 9, 12, 14, 16, 17, 18, 21, 25, 27 and 28 of the *Water Resource Management Method in Xi'an City*	A	Environmental management program and operational control procedure
5	Casting smoke and dust	Smoke and dust	Casting branch	Smelting and cupola discharge	Air pollution	Now	Normal	Article 2 of section IV of the *Emission Standard of Air Pollutants of Industrial Furnace of People's Republic of China*	B	Environmental management program and operational control procedure

Source Dang [1]

Table 14.6 Environmental objectives, indicators, and actions of a specific machine manufacturing company in 2006

No.	Environmental aspect	Objective	Indicator	Environmental management program
1	Engine oil leakage	Free from land pollution	Land pollution free	The equipment maintenance shop should be supervised by the product department and environmental security office to regularly check, evaluate, and score equipment for oil impregnate/leakage The equipment maintenance shop should perform regular checks of the equipment in the entire company, once problems are identified The equipment repaired in 2005 should be rechecked, and any problems found reported
2	Water/power resource consumption	Reduction of water and power consumption	Compared with 2005, water consumption decreased 10,000 t and the power consumption remained the same as the previous year	Periodic monitoring should be performed of the water condition of the line frequency furnace, resin furnace, and air compression station. Recirculation water is constantly produced to improve the ability of recycling and processing wastewater (1000 t) Monitor and maintain the hydrostatic test system of the drying cylinder to improve the ability of recycling wastewater (2000 t) Water metering is strengthened and perfected to ensure the accuracy and rationality of water use data (1000 t) Water-saving technology improvement is updated in the all water facilities in the boiler (3000 t) Power consumption management is strengthened, the reconstruction of electricity facilities is accelerated, and efforts are made to improve power-consuming efficiency, to further perfect the electrometering equipment and electric cost of contribution and to reduce electricity waste. Statistics and issuing are implemented for the water/power consumption situation of each department by the power office, to ensure fire and power are controlled

(continued)

Table 14.6 (continued)

No.	Environmental aspect	Objective	Indicator	Environmental management program
3	Dust emission of casting cupola	Smoke and dust emission compliance	Smoke density is 200 mg/m^3 Exhaust blackness Ringelmann-level: 1	To remold the cupola dust removal installation, employ secondary dedusting methods for the high-temperature flue gas, lower the smoke and dust temperature to the first stage to remove particle dust, and employ a bag filter during the second stage

Source Dang [1]

(III) Example 3: resource, function, responsibility and authority

In ISO 14001, the clause 4.4.1 regulations concerning the resource, function, responsibility, and authority are as follows:

The manager should ensure essential resources are provided for the establishment, implementation, maintenance, and improvement of the environmental management system. Resources include special skills for HR, infrastructure, and technical and financial resources of the organization.

For the effective implementation of the environmental management activities, the function, responsibility, and authority should be explicitly stipulated within a document for delivery.

A professional management representative should be assigned by the chief executive of the organization; regardless of other aspects of his or her responsibilities, his or her function, responsibility, and authority should be explicitly stipulated, so that an EMS can be established, implemented, and maintained in accordance with the requirements of the standards. The performance of the EMS is reported to the chief executive, and suggestions for improvement are proposed.

For the manufacturing shop, an assessor should understand ① the specific product/activity produced by the workshop; ② the organization, staff composition, and corresponding responsibilities of the workshop; ③ the infrastructure, production and processing equipment, and environmentally protective facility of the workshop; and ④ the environmental aspect of the workshop, including its primary environmental aspect and its performance control condition.

(IV) Example 4: contingency reserve and response

Requirements of the contingency reserve and response are stipulated in clause 4.4.7 of ISO 14001. One or more measures should be established, implemented, and possessed to identify the situation and accident possibly causing emergency environmental impact; response measures should be stipulated. The organization should respond to emergency situations and accidents and prevent or reduce their damaging environmental impacts. The organization should regularly review, test, revise, and perfect their contingency reserves and response measures.

For example, the contingency plan prepared by a specific machinery manufacturing company includes the following contingency plans: ① fire prevention; ② poison gas and acetylene explosion prevention; ③ oils and dangerous chemical leakage pollution (prepared by different dangerous chemical MSDS); ④ emergency sewage treatment facility; and ⑤ emergency at an important, environmentally protective facility.

(V) Example 5: record of environmental governance process

It is stipulated in clause 4.5.4 of ISO 14001 that based on its needs, an organization should establish and maintain necessary records for verifying conformity of EMS and the standard requirements and their achieved results. One or more procedures should be established, implemented, and maintained by the organization for the recording logo, storage, protection, retrieval, preservation, and disposal. The environment record should be legible, identifiable, and possess traceability.

For example, the record of a specific company's environmental governance process includes the following:

(1) Complaint record;
(2) Training record;
(3) Process inspection record;
(4) Checking, maintenance, and correction record;
(5) Record of related supplier and contracting party;
(6) Record of accidental event;
(7) Contingency reserve trial sheet;
(8) Audit result;
(9) Management review result;
(10) External decision concerning the exchange of information;
(11) Record of environmental laws and regulations requirement;
(12) Record of important environmental aspect;
(13) Environment conference record;
(14) Environmental performance information;
(15) Record of laws and regulations conformity; and
(16) Exchange with related party.

14.4 What Are the Improvement Measures?

The measures for improving enterprise behavior are discussed from the aspect of an enterprise's environmental management measures. These include green product certification, extension of entrepreneurs' responsibility, green marketing, and establishment of an environmental management system.

14.4.1 Green Product Certification

(I) Green product and environmental label

Green product refers to a type of product whose production process has the properties of energy saving, water saving, low pollution, low toxicity, regeneration, and recovery. A green product is in accordance with the principle of sustainable development, and it is also the final embodiment of a green technological application.

To promote the wide application of green products, the certification management of green products is stipulated by the government. In terms of the conformity to the sustainable development principle, for a green product finishing its production

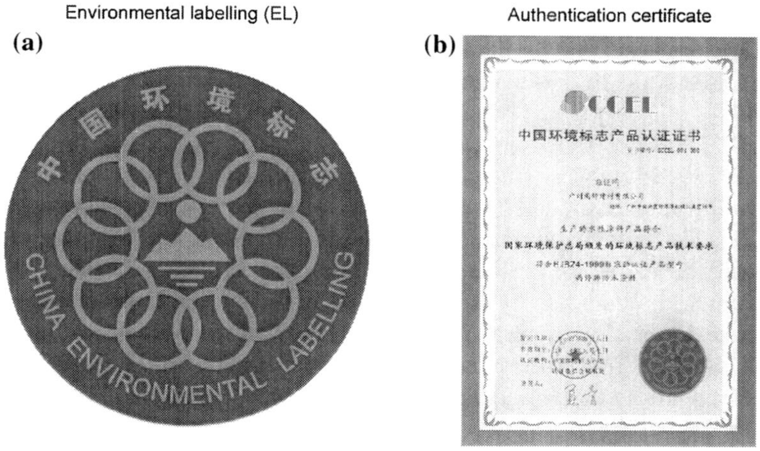

Fig. 14.1 Example of China's environmental labeling and authentication certificate

process according to a specific mode of production, after being certified by the national specialized agency for product management, it can be granted "EL," and concurrently, its manufacturing enterprise can obtain the "authentication certificate of an environmental label product."

"Green product" and "EL" are closely related; an EL is the outcome of a green product certification. Here, EL is also called eco-label, green label, and environment label. It is a label attached to a product and issued by the government's environmental management department to some commodities according to relevant regulations and standards. It is used to identify that a product is in accordance with the stipulated environmental protection requirements during the entire process from production to use and recovery. The product is harmless or causes only slight harm to the ecological environment and is convenient for resource recovery and recycling. Figure 14.1 is an example of the Chinese environmental labeling and product authentication certificate of a specific enterprise. In application, whether a product is a "green product" can be identified by observing whether the product is marked with "EL."

(II) EL system

The EL system is an important means for environmental management of governmental products. It is primarily embodied in the implementation of EL certification of an enterprise's products. This system requirement confirms the product's environmental performance and production process (including the entire process from product design, production, use and waste processing and disposal, and even recycling and reusing) by the institution or non-governmental organization designated by the country in accordance with the environmental product standards (also called technical requirements) and relevant regulations. Consumers are informed which products conform to environmental protection requirements via a logo, which is more beneficial for the ecological environment.

The bases of the EL system are the international standard ISO 14020 EL series, ISO 14001, ISO 14004 and domestic standards GB/T 24020 series, GB/T 24001, and GB/T 24004, as shown in Table 14.2.

The EL system primarily involves three core elements: certifications' executive body, certification basis, and EL distribution. In our country, the executive body of the EL system is the environmental certification center of the Ministry of Environmental Protection (hereinafter referred to as the certification center). It is the institution that is authorized by the Ministry of Environmental Protection to perform authorized certification of green products on behalf of the country and to award a product the EL. The primary basis for the certification of EL is the relevant international and national standards concerning product quality and EL, as shown in Table 13.2 of Chap. 13, namely an example of an EL certified product and its applicable standard. Finally, EL is issued to the product conforming to the green product standards, which is not only the affirmation of product's environmental performance, but is the enterprise's objective and outcome of developing green products.

The ultimate purpose of issuing EL is to protect the environment, and it is embodied in the following two aspects: to send a message to consumers via an EL label to notify them which products are beneficial for the environment and to guide them to purchase and use these products. Second is to guide an enterprise to consciously adjust product structure and use cleaner production techniques via consumer's selection and market competition, to ensure the enterprise's environment protection behavior obeys laws and regulations and produces environment-benefiting products.

14.4.2 Extended Producer Responsibility (EPR)

(I) Concept of EPR

To promote the overall improvement of the environment, an enterprise's responsibility is extended to product maintenance or scrap recycling and waste disposal from the original single product design and production responsibility; this is called extended producer responsibility (EPR). By means of the EPR system, an enterprise is requested to assume the full responsibility for the life cycle and environmental protection of its own products and to fulfill the obligation of environmental protection while serving society, especially in the phases of product recovery, recycling, and final treatment.

(II) EPR subject of liability and responsibility assignment

As in the product life cycle, it needs to change producer, transportation staff, users, and other multiple managers; therefore, the EPR subject of liability changes continuously and primarily includes the producer, seller, consumer, and government. Different subjects of liability bear different responsibilities.

EPR includes the following: ① The responsibility to produce innocuous and harmless products—It is stated in the *Instruction for Promoting Green Consumption* jointly issued by the State Development and Reform Commission, CPC Propaganda Department, and another department that "EPR is established and perfected to encourage a manufacturing enterprise to minimize the utilization of toxic, harmful, degradation-resistant, disposal-resistant, and strong, volatile substances". ② To be responsible for product recovery and use—For this responsibility, the procedure can establish a special product recovery and regeneration department to complete the waste recycling, and the responsibility can be assigned to a third party; e.g., scrapped product is recovered by a retailer, and the producer is only responsible for recycling. ③ Information responsibility—The producer is obligated to specify the commodity material, recovery method, and other matters of the product instruction or product packaging. It is stipulated in the *Instruction for Promoting Green Consumption* that an enterprise should actively disclose its energy efficiency, water efficiency, environmental performance, carbon emission, and other information concerning products and services and to encourage the implementation of self-declaration, publicity, and supervision system of the product standard. ④ To share the recovery and disposal cost of the waste product—This can be determined based on the waste disposal cost, speed, and processing ability of the recycling enterprise and product output of the producer; it is allocated between the producer and the reclaimer by a ratio.

The seller should assume the relevant responsibilities of recovering waste products, charging selling expenses, returning deposits, and choosing and storing recycled waste products. Additionally, the seller has an obligation to disclose the environmental message of the products to be sold and, in the process of promoting products, to provide relevant product information, return deposits, and other matters to the consumer.

The consumer responsibility is to first deliver the waste products to the recycling point or designated site and, second, to share the recovery processing cost of the waste products. Usually, there are three ways of paying the consumer: The first is a down payment; namely, the processing costs have been added to the product price while purchasing products; the second is a discarding fee; namely, a specific amount of money is paid when product is discarded by the consumer; this type of mode can encourage consumers to extend product service life and reduce discharging quantity but may also cause the problem of improper disposal; and the third is a deposit, which is widely used in beverage bottles, batteries, and tires.

As the constitutor and source of the EPR system, the responsibility of the government is to prepare the EPR legal system and relevant technical regulations, including the technical specifications for product classification standards, scrapping standards, and recovery disassembly. Preparation of EPR policy includes the execution of a government green procurement and green consumption policy. The enterprise performance assessment system is established with the EPR executive condition as an important factor of assessment, and EPR supervision is implemented.

(III) Example: development status of China's EPR system

In 2007, the *Recovery Regulations of Renewable Resources Act* was issued and implemented. It expressly stipulates that manufacturing enterprises should sell their productive waste metal by signing a purchase contract with a renewable resource recovery enterprise to indicate that the EPR system is incorporated into the waste resource management system in China. In 2008, the EPR system was officially written in the *Circular Economy Promotion Act of the People's Republic of China.*

It is expressly stipulated in Article 15 of the *Circular Economy Promotion Act of the People's Republic of China* that: "the enterprise whose product or packaging is listed in the compulsory recycling directory must recycle the relevant abandoned products or packaging. If available, it is utilized by its own manufacturing enterprise; if it is unavailable due to lack of technical and economic conditions, the manufacturing enterprise is responsible for its harmless disposal." If the seller or another organization is entrusted by the producer to recycle waste products or packaging or a waste recovery or disposal enterprise is entrusted to perform recycling, the commissioned party should be responsible for recovery, utilization, or disposal in accordance with relevant laws, administrative regulations, and agreements of contract.

It is not difficult to observe that according to the *Circular Economy Promotion Act of the People's Republic of China*, the implementing object of EPR is the compulsively recovered product or packaging. In February 2009, the *Recycling Control Regulations on Waste Electrical & Electronic Products Act* was issued by the State Council to bring the recovery processing of electronic wastes under the legal system's purview. In 2010, "four machine and one computer" (i.e., television set, refrigerator, washing machine, and air conditioner and computer) were listed in the first version of the processing directory. In 2015, the *Processing Directory of Waste Electrical & Electronic Products (Version 2014)* was issued. The new directory includes mobile communication handsets, fax machines, copying machines, and other products to further expand the implementation scope of the *Recycling Control Regulations on Waste Electrical & Electronic Products Act*. It is stipulated in the *Overall Reform Scheme of Eco-civilization System Act* printed and issued by the State Council in 2015 that compulsory recovery is implemented for composite package, battery, agricultural film, and other low-value wastes.

When EPR is implemented in our country, we should perform specific analysis of specific issues based on different products or packaging to ensure the environmental and economic benefits of the EPR system implementation. EPR should be imposed on the wastes with significant environmental hazard according to legal provisions; additionally, suitable economic means should be applied to assign proper responsibilities and obligations to the relevant subject. For the waste with minor environmental hazard and high recycling value, the producer or other social community can be guided to recycle and reuse it by voluntary means or market mechanisms. For example, multichannel recovery, fund subsidies, centralized processing, and other principles are stipulated in the *Recycling Control Regulations on Waste Electrical & Electronic Products Act* issued by the State Council in 2009 and in the Regulations on the Collection and Application of the Processing Funds of Waste Electrical & Electronic Products Act issued in 2012. The producer of the electrical and electronic product (including the consignee or its agent of the imported electrical and electronic product) is requested to capture funds for the subsidy of the recovery processing cost of the waste electrical and electronic products. The implementation of the fund subsidy system effectively promotes the electrical and electronic product manufacturing enterprise to participate in the electronic waste recovery; e.g., the electronic waste recovery processing networks have been built by Gree, Changhong, TCL, and other manufacturing enterprises to significantly promote the processing capacity of waste electronic products.

14.4.3 Green Marketing

(I) Concept of green marketing

Green marketing refers to the activities implemented by the enterprise to meet consumers' green demands and to perform environmental protection via the market exchange process of the product. It includes pricing, distribution, after-sales service, and other business activities.

During the implementation of green marketing and guided by the market requirements with green marketing as the center and with the purpose of enterprise profit and environmental protection, the enterprise should carefully conceive, elaborately design, and perform clean manufacturing of its own products and should consider the sales of product recovery after scrapping, to integrate their self-interests, consumer benefits, and environmental protection benefits.

(II) Characteristics of green marketing

Green marketing is an extension and expansion of traditional marketing, but it has a unique and fresh new connotation in marketing concepts and marketing mix strategies. It has the following characteristics:

(1) The marketing has a distinctive "green" mark. In an entire market survey, product development, pricing, distribution, and after-sales service, it is closely related to the "green" concepts of safeguarding ecological balance, valuing environmental protection, and improving people's life quality, throughout the entire marketing activity.
(2) Comprehensive feature. Ecological and societal marketing concepts are integrated by green marketing to promote the enterprise to further provide consumer benefit and people's common aspiration; it represents the mechanism for an enterprise's survival, development, and trend.
(3) The basis of green marketing is the consumer's enthusiasm for "green" consciousness. Based on the consumer's psychology to pursue improvement in their quality of life and their environment, an appeal is made to their

reminiscent mood and desire to return to innocence and stimulate consumer's demands for a healthy, safe, and clean environment, to produce the green industry, product, and marketing.

(4) The purpose of green marketing is to achieve sustainable use of resources and to protect and improve the ecological environment. The producer is requested by green marketing to constantly reduce the utilization of non-renewable resources and to develop additionally renewable resources, to prevent environmental pollution and maintain an ecological balance.

(5) The basis of green marketing are the green product and industry. The utilization and production of green products can encourage the development of natural resources, R&D of the raw material industry, technique and process equipment, product sales and use, and other different divisions, to form a compound industry based on "green" techniques. Generally, an enterprise is engaged in one or more products' production and sales in a green industry.

(6) Green standards and signs represent world indifference. There are different standards for green products in various countries, but they require that product quality, production, utilization, consumption, disposal, and other aspects should be in accordance with the requirements of environmental protection and support the ecological environment and human health.

(7) Characteristics of green marketing methods. Principally, by imitating the coordination mechanism of nature, an enterprise represents the integrative development of human behavior and nature, the so-called unity of heaven and man, to build an enterprise's green image and to promote product sales and enterprise development.

(III) Market entity and behavior of green marketing

The producer and consumer are the market entities of green marketing. The next sections contain a brief analysis of the behaviors of these two market entities from an economics perspective.

1. Producer

A producer is the market entity, and there are three tasks in green marketing: green product production, green product marketing, and environmental quality protection. The producer also produces by-products while manufacturing products, e.g., wastewater, waste residue, and exhaust. If these by-products cannot be used second time, they are expensive to process or they would pollute the environment. Therefore, a producer should ensure the use of by-products as an input, as using by-products as a portion of input can reduce the cost. However, many producers do not use environmentally friendly by-products, causing too much environmental debt and resulting in environmental pollution and environment quality deterioration. The purpose of green marketing is to expect the producer to implement the concept of environmental protection in product production and marketing; with the development of the enterprise, the environmental quality improves gradually.

2. Consumer

In green marketing, the consumer has dual tasks: to pursue green and self-disciplined consumption to reduce his or her environmental impact and to share the enterprise's economic input for the improvement of environmental quality. This is because environmental quality is closely related to the consumer. Human health will be influenced by polluted air, deteriorating water quality, and environmental noise; additionally, productivity loss will be caused by environmental pollution, e.g., a price upswing caused by a reduction and undersupply of food supplies, resulting in the direct loss of personal benefit. Therefore, to maintain other conditions, there is a proportional relation between consumer benefit and environmental quality. Improvement of environmental quality will be beneficial to improve consumers' benefit level. A green product is purchased by consumers to indirectly promote the improvement of environmental quality and assist ourselves to improve the environmental benefit level.

From the above analysis, it can be observed that both producer and consumer can fulfill their responsibility to environmental protection via green marketing and can provide the economic input for environmental improvement, to strongly encourage the implementation of an environmental protection strategy.

In addition, the government has a leading and guarantee role in green marketing. For example, green product consumption is clearly encouraged in the *Instruction for Promoting Green Consumption Act* issued in 2016, which requests that high-efficiency and energy-saving machines, energy conservation and environmental protection automobiles, efficient lighting products, and other energy-saving products are continually promoted. By 2020, the market share of air conditioners, refrigerators, water heaters, and other energy-saving household appliances with an energy efficiency label greater than two will reach more than 50%. We must strengthen the promotion of the new energy automobile, accelerate the construction of an electric car charging infrastructure, and encourage the indispensable management role.

14.4.4 Establishment of an Environmental Management System

A system is a generic term for the regulations, rules, norms, and methods that should be jointly followed by the members of a social organization or group. To standardize the enterprise's environmental behavior, our country has formed multiple systems, such as environmental impact assessment and pollutant discharge license control. Several systems are selected for brief introduction in the following subsections.

(I) Environmental impact assessment system

The environmental impact assessment system is specific to engineering construction that may cause an environmental impact. A survey, prediction, and evaluation of the environmental impact of the surrounding area that may be caused by the activity are performed before planning, a policy for preventing environmental pollution and damage is proposed, and corresponding solutions are prepared. The environmental impact assessment system is the major legal means to achieve the synchronous development of economic construction, town and country construction, and environmental construction. It not only must perform economic evaluation but must also perform environmental impact assessment of a construction project, to scientifically analyze possible environmental problems in the construction activity and to propose controlling measures.

The implementation of an environmental impact assessment is usually subject to the following procedures. ① The developer must entrust a specialized agency that is qualified to perform environmental impact assessments for environmental survey and integrated forecasting; this agency develops an environmental impact report. ② The developer must publish the report and listen carefully to the public's and experts' opinions. For different opinions, it is stipulated by some countries that a public opinion hearing must be held. ③ Based on the opinions of the public and the experts, any necessary amendment is made to the solution. ④ A department is responsible for the final examination and approval to determine whether the engineering construction is allowed.

An environmental impact assessment provides the basis for the national site selection of the construction project to prevent environmental damage that will be difficult to eliminate being caused by a non-reasonable layout. Based on the environmental impact assessment, the current surroundings can be clarified, the scope, degree, and trend of the environmental impact of the construction project can be forecast, and the targeted environmental protection measures can be proposed. Furthermore, the environmental impact assessment also can provide the scientific basis for the environmental management of the construction project.

The *Law on Environment Impact Assessment* was implemented in September 2003 to stipulate the regulations concerning environmental impact assessments of planned projects and construction projects in detail and to prepare the legal liability to be assumed by the preparation authority, approving authority and project construction organization.

It is stipulated in Article 19 of the *Environmental Protection Law of the People's Republic of China* implemented in January 2015 that "environmental impact assessment should be implemented according to laws while preparing the relevant development and utilization planning, and building the project that has an impact on the environment. The development and utilization planning whose environmental impact assessment is not implemented according to laws cannot be implemented, and construction projects whose environmental impact assessment is not implemented according to laws are not allowed to begin building." In the situation when construction is halted because the environmental impact assessment is not implemented and the organization refuses to implement the environmental impact assessment, it is stipulated in Article 63 that "the case is transferred to the public security organization by the responsible department for environmental protection of the people's government above the county level or other relevant department, and the direct governor and other directly responsible personnel should be punished with 10–15 days of detention; for minor cases, the punishment is five to ten days of detention."

(II) Environmental protection liability system

The environmental protection liability system is the system that requests the enterprise and public institution discharging pollutants assign responsibility for pollutant emission, the environmental protection post, and other relevant staff. By means of such a system, the enterprise, public institution, and other production operators discharging pollutants are urged and supervised to employ effective measures to prevent exhaust gas, wastewater, waste residue, medical waste, dust, foul gas, radioactive substance, noise, vibration, optical radiation, and electromagnetic radiation produced in its production construction or other activity and to avoid the pollution and damage to the environment caused by these pollutants. An environmental protection liability system has been implemented in our country's air pollution control, water pollution control, and other many fields. It is stipulated in Article 42 of the *Environmental Protection Law of the People's Republic of China* implemented in January 2015 that the enterprise and public institution and other production operators that discharge exhaust gas, wastewater, waste residue, medical waste, dust, and other pollutants "should establish an environmental protection liability system, to clarify the responsibility of the unit head and relevant personnel." It is further stipulated that "monitoring equipment should be installed by the key pollutant discharging unit according to the relevant state regulations and monitoring provisions, to ensure the normal operation of equipment and to capture the original monitoring record. It is prohibited to discharge pollutants by escaping supervision via closed conduit, seepage well, sewage pit, perfusion or falsified/false monitoring data, or legal operation of the pollution-prevention facility."

(III) "Three Simultaneities" system

A "three simultaneities" system specifies that for all newly built, rebuilt, and expanded basic construction projects (including small projects), technological transformation projects, natural development projects, and other engineering projects possibly causing damage to the environment, facilities for preventing pollution and other public nuisances and other environmental protection measures must be designed, built, and placed into production concurrently.

The "three simultaneities" system is the first environmental management system issued by our country; it was released in the *Several Regulations on Environment*

Protection and Improvement approved by the State Council in 1973. It was incorporated into the *Environmental Protection Law of the People's Republic of China (Trial)* issued in 1979. It is stipulated in Article 41 of the *Environmental Protection Law* implemented in January 2015 that "in a construction project, a facility for preventing pollution should be designed, built and placed into use with the construction activities. The facility for preventing pollution should be in accordance with the requirements of the approved environmental impact assessment documents, and it shall be not arbitrarily dismantled or idled." The "three simultaneities" system is incorporated throughout a planning and construction project; it is a safeguard measure before an event in the entire process from feasibility study, preliminary design, construction, and pilot production to final completion acceptance.

(IV) Pollutants discharge license management system

The pollutant discharge license management system refers to the system that any enterprise and public institution or production operator that must discharge pollutants into the environment must apply to the environmental protection department for approval; after obtaining sewage discharge permission, the organization is allowed to discharge pollutants.

The pollutant discharge license system is an internationally accepted environmental management system. The *Regulation on the Discharging License of Key Water Pollutants in Huaihe River and Taihu Basins (Trial)* issued by the original State Environmental Protection Administration in 2001 is China's first administrative provision concerning pollutant discharge licenses for the drainage basin. Then, the pollutant discharge license system is explicitly stipulated in the *Water Pollution Prevention Law*, *Atmospheric Pollution Prevention Law* and other environmental protection laws and regulations. It is stipulated in Article 45 of the *Environmental Protection Law* implemented in January 2015 that "the enterprises and institutions and other production operators implementing pollutant discharge license control should discharge pollutants according to the requirements of their pollutant discharge license, and those without a pollutant discharge license are not allowed to discharge pollutants." It is further stipulated in Article 55 that "the key pollutant discharging unit should truthfully disclose the name of the major pollutants, discharging methods, emission concentration, total quantity, excessive emission situation, construction and operation conditions of the facility for preventing pollution to the public and should receive social supervision." In addition, for the enterprise in violation of the legal provision, it is further stipulated in Article 63 that for the enterprise and public institution or production operator that are instructed to halt construction but refuse to halt as pollutants are discharged without a pollutant discharge license, "in addition to paying fines in accordance with relevant laws and regulations, the case is transferred to the public security organization by the responsible department for environmental protection of the people's government above the county level or other relevant department, and the direct governor and other direct responsible personnel should be punished with ten to fifteen days of detention; for minor cases, the punishment is five to ten days of detention."

Accurate verification of the enterprise's emission amount is the core work of license management. Our country's Ministry of Environmental Protection has established a complete set of excellent verifying and accounting methods for an enterprise's total pollutant amount, and the operational supervision systems for a pollution control facility. The complete emission reduction verification is comprehensively implemented in the key pollution source of the "six factory (site) and one vehicle" (i.e., thermal power plant, cement plant, iron and steel plant, sewage treatment plant, paper mill, livestock, and poultry farm and motor vehicle), and the emission amounts are verified according to the performance of the pollutant control facility and resource energy consumption, so that we can make efforts to understand the actual pollution discharge of the enterprise more accurately, resulting in positive achievements. Currently, our country is beginning to attempt a "one certificate" pollution regulation; namely, with the pollutant discharge license system as its core, the "certificate-carrying" content of the pollutant discharge license is applied to support the effective connection of a pollutant discharge license and environmental standards, environment monitoring, environmental statistics, environmental impact assessment, total quantity control, pollution charges (environment tax), license monitoring, and other systems. The unique directory of the stationary pollution source information is established for the code management of the sources of pollution, to achieve integration, to share the pollution sources emission information, and to effectively encourage cooperative governance.

In addition, the total quantity control system, environmental monitoring system, environmental protection goal responsibility system, and examination and evaluation system are further stipulated in our country's environmental protection act. These systems are critical for the integrated environment management of the enterprise.

Classroom Discussions and Assignments

Topic Discussion

The requirement for enterprise environmental improvement is discussed in this chapter from the perspective of business management. It includes the quality of the manager, management content, scope, and basic management methods and measures. In recent years, many environmental pollution events have occurred in our country that reflect many disadvantages of an enterprise's environmental management. Please choose a typical environmental pollution event in our country and, based the course content, discuss the cause of the event in a group. Propose

Table 14.7 Grading of the group report in class

Report title	Logic (maximum 10 points)	Ability to apply integrated knowledge (maximum 10 points)	Innovation (maximum 10 points)	Presentation skill (maximum 10 points)	Total points

measures or suggestions for improving an enterprise's environmental management. List the research outline of the relevant topics.

Homework and Presentation in Class

In the previous chapter, we requested students leave the classroom and perform a field survey based on actual situations, to comprehensively apply what they have learned in the classroom to evaluate and solve problems. The task of this lesson is to continuously exercise students' comprehensive and problem-solving abilities; we hope that students can comprehensively apply what they have learned in the entire course to propose their own innovative thinking for the to-be-solved problems. The task of this lesson is to execute a survey based on a typical environmental pollution event selected in the class discussion and the research outline to develop an approximately 3000-word research report, to be shared in class, using PowerPoint, via the group. The grading scale is shown in Table 14.7.

Pre-graduation Thesis of the Curriculum

As a general education curriculum for undergraduate freshmen, guided by sustainable development and focusing on the circular economy, more than ten course topics have been discussed concerning the "Circular Economy and Sustainable Development Enterprise," with the goal of guiding students to learn methods to propose problems, methods to perform research, and methods to settle and report the research results, which cultivates academic talent. For the completion of this course, students are required to choose any of the topics discussed during the course, independently determine a title, and write an approximately 3000-word paper. It is requested that the paper comprehensively reflects the overall process of scientific research, integrated application, and innovation ability of the curriculum knowledge. For the paper format, please reference the format requirements of manuscripts in the periodical *Environmental Science*. The grading scale is shown in Table 14.8. The pre-graduation thesis of the curriculum is an individual paper, accounting for 50% of the total points.

Table 14.8 Grading of the group report in class

Report title	Process of topic selection and research basis (maximum 10 points)	Research method, result, and conclusion (maximum 10 points)	Integrated application of curriculum knowledge (maximum 10 points)	Expression and specification (maximum 10 points)	Total points

Recommended and Referenced Literatures

1. Jixiang, Dang. 2007. *Understanding and implementing of ISO14001 environment management system standard*. Xi'an: Northwest University Industrial Publication.
2. National Development and Reform Commission, The Propaganda Department of the Central Committee of the CPC, Ministry of Science and Technology, etc. 2016. Instruction for promoting green consumption: No. 353 F.G.H.Z. [EB/OL]. (2016-06-20) [2016-08-09]. http://www.mep.gov.cn/gzfw_13107/zcfg/hjjzc/gjfbdjjzcx/qtjjzc/201606/t20160623_355542.shtml.
3. The State Council. Overall reform scheme of eco-civilization system [EB/OL]. (2015-09-22) [2016-08-09], http://www.mep.gov.cn/xxgk/hjyw/201509/t20150922_310133.shtml.
4. Changchun, Hua. 2010. *Cleaner production technology and application of industrial enterprise*. Nanjing: Jiangsu Science and Technology Press.
5. Rentian, Ji, and Yu. Junlin. 2007. *Understanding and implementing of ISO 14001: 2004 standard*. Beijing: China Metrology Press.
6. Yanping, Li. 2005. Discussion on the extended producer responsibility system. *Environmental Protection* 7: 13–18.
7. Ying, Liu, Zhang Genbao, and Zhang Dan. 2001. Total environmental management and cleaner production. *Jiangsu Machine Building & Automation* 94: 10–13.
8. Bangsheng, Luo, and Li Anzhou. 2005. *Corporate environmental management*. Chongqing: Chongqing University Press.
9. Xiaoming, Mao, Jiao Zhifeng, and Liu Weiming. 2010. *Regulations and standards manual of environment management system*, 2nd ed. Beijing: China Standards Press.
10. Min, Qi. 2008. Improving enterprise manager quality and adapting the requirement of external environment. *Journal of Liaoning Educational Administration Institute* 3: 113–115.
11. Yong, Qian. 2004. Extended producer responsibility policy in OECD member countries and its effect on market structure and firm conduct. *Industrial Economics Research* 2: 9–18.
12. Wenju, Ren, and Li Zhong. 2006. Theories and practices of extended producer responsibility system. *Economist* 4: 30.
13. Houfen, Wan. 2001. *Green marketing*. Beijing: Higher Education Press.
14. Gan, Wang. 2006. Improving the extensive manufacturer responsibility regime in China. *Modern Law Science* 4: 167–173.
15. Jingwei, Wang. 2005. Induction of EPR system of EU waste electronic and electrical equipment. *Appliance Technology* 1: 35.
16. Lei, Wang, and Liu Ya. 2009. *Application examples of the latest standards of the environment management system*. Beijing: Chemical Industry Press.
17. Minhua, Wang. 2009. *Management system and certification*. Beijing: China Metrology Press.
18. Zhaohua, Wang, and Wang Jianhua. 2006. On E-waste management based on extended producer responsibility. *Journal of Beijing Institute of Technology (Social Science Edition)* 8: 49–51.

19. Subin, Wen, and Xue Hengxin. 2005. Sustainable development-oriented extended producer responsibility system. *Economic Problems* 2: 11–14.
20. China Quality Certification Center, Hefei University of Technology. 2009. *Green assessment method and application of consumable products*. Beijing: China Standards Press.
21. Dingchun, Zou. 2011. *Interpretation and practice of common environmental protection laws*. Beijing: China Environmental Science Press.

CPSIA information can be obtained
at www.ICGtesting.com
Printed in the USA
LVOW10*1615020418
571979LV00008B/243/P

9 789811 085239